Library of
Davidson College

A Systems View of Man

Also of Interest

Autopoiesis, Dissipative Structures, and Spontaneous Social Orders, edited by Milan Zeleny

Sociobiology: Beyond Nature/Nurture? Reports, Definitions, and Debates, edited by George W. Barlow and James Silverberg

Assessing the Contributions of the Social Sciences to Health, edited by M. Harvey Brenner, Anne Mooney, and Thomas J. Nagy

Sign Language and Language Acquisition in Man and Ape: New Dimensions in Comparative Pedolinguistics, edited by Fred C. C. Peng

Psychology and Education of Gifted Children, Philip Vernon, Georgina Adamson, and Dorothy Vernon

Health Care for the Whole Person: The Complete Guide to Holistic Medicine, edited by Arthur C. Hastings, James Fadiman, and James S. Gordon

Ethics in an Age of Pervasive Technology, edited by Melvin Kranzberg

Goodbye to Excellence: A Critical Look at Minimum Competency Testing, Mitchell Lazarus

About the Book and Author

A Systems View of Man
Ludwig von Bertalanffy
edited by Paul A. LaViolette

What does it mean to be human? What distinguishes man from other animals? "Man's creation of the universe of symbols," replies Ludwig von Bertalanffy. "Man lives in a world not of things, but of symbols." Dr. von Bertalanffy explores the historical development of symbolic language, examines the nature of human values, and shows how a current breakdown of symbolic universes contributes to the feeling of meaninglessness so prevalent in modern society. He notes that a major portion of mankind's aggressive acts are not biologically induced but arise within symbolic frameworks.

Dr. Ludwig von Bertalanffy, generally considered the father of general systems theory, applies a systems model to human functions and uses that model to explain aspects of the central nervous system, creativity, and mental illness. He also shows how many practices in psychology and education serve to stifle individual creativity and development. Originally a biologist, the late Dr. von Bertalanffy is best known as a humanitarian and general systems theorist who applied interdisciplinary perspectives to a wide spectrum of topics. His essays in this book draw from anthropology, linguistics, psychology, neurophysiology, sociology, and philosophy.

Paul A. LaViolette, editor of this collection of essays, is at Portland State University.

A Systems View of Man

Ludwig von Bertalanffy

edited by
Paul A. LaViolette

Westview Press / Boulder, Colorado

All rights reserved. No part of this publication may be reproduced or transmitted in any form or by any means, electronic or mechanical, including photocopy, recording, or any information storage and retrieval system, without permission in writing from the publisher.

Copyright © 1981 by Maria von Bertalanffy

Published in 1981 in the United States of America by
 Westview Press, Inc.
 5500 Central Avenue
 Boulder, Colorado 80301
 Frederick A. Praeger, Publisher

Library of Congress Cataloging in Publication Data
Bertalanffy, Ludwig von, 1901-1972.
 A systems view of man.
 Bibliography: p. 165
 Includes index.
 1. Man—Addresses, essays, lectures. 2. Symbolism—Addresses, essays, lectures. 3. Mind and body—Addresses, essays, lectures. 4. System theory—Addresses, essays, lectures. 5. Psychology—Methodology—Addresses, essays, lectures. I. LaViolette, Paul A. II. Title.
BD450.B326   128   80-20803
ISBN 0-86531-084-X
ISBN 0-86531-094-7 (pbk.)

Printed and bound in the United States of America

Contents

Preface ix
Source Notes xi
Introduction xv

1   Man's Universe of Symbols 1
2   Human Values in a Changing World 9
3   Comments on Aggression 23
4   Systems Perspectives on the
     Problem of Mental Illness 31
5   A Definition of the Symbol 41
6   An Etymology of Symbolism 57
7   The Evolutionary Origins of Symbolism 69
8   The Mind-Body Problem:
     A New View 85
9   General Theory of Systems:
     Application to Psychology 109
10 Toward a Generalized Theoretical
     Model for Psychology 121
11 Problems of Education in America 133
12 On Interdisciplinary Study 145

Notes 159
Bibliography 165
Appendix 173
Index 177

Preface

According to system theory, the difference between a "collection" and a "system" is that in a collection the parts remain individually unchanged whether they be isolated or together, i.e., they are simply a sum, whereas in a system the parts necessarily become changed by their mutual association; hence, their whole becomes more than just the sum of the parts. Editing a group of essays provides an interesting illustration of this principle. In order to transform a collection of essays to read as a book, the individual works, besides being properly ordered, must undergo certain internal revisions to make them compatible with one another. In other words, the essays must be made to hang together as a system.

It was in this spirit that the editing of this book was carried out. Several revisions were necessary. For example, the latter portion of the essay "A Biologist Looks at Human Nature" (1956) was adapted to serve as an introduction in Chapter 1 to the topic of symbolism. The first portion of the essay, concerned with the biological uniqueness of the human species, was eliminated, and a more relevant portion concerned with the evolution of the human brain was made into an appendix. These sections, which were an appropriate beginning for the essay by itself, would have been misleading if placed as an introduction to the book, whose main theme at the start centers on symbolism.

Another essay, "The World of Science and the World of Value" (1964), was divided into two portions. The first part, dealing with modern society and human values, was integrated with similar material in the essay "Human Values in a Changing

World" (1959) to form Chapter 2, while the second part, dealing with the shortcomings of the American educational system, appears separately as Chapter 11. The essay "On the Definition of the Symbol" (1965), which separated naturally into three parts, was used to form chapters 5, 6, and 7. Also, a short essay entitled "Reconsideration 1972: A Mini-History of the Concept of Symbolism" was integrated into Chapter 6.

In several instances a paragraph from one essay was inserted into another essay where it seemed more appropriate. Other editorial revisions included the elimination of repetition, minor stylistic improvements, and some suitable rearrangements of sections within essays to provide a smooth transition from one chapter to another. All the while, however, care was taken to preserve unchanged as much as possible of the original texts. Those wishing to study the essays individually in their original form may consult the Source Notes for proper references.

I would like to thank Maria von Bertalanffy, Ervin Laszlo, Tom Doulis, and Jerzy Wojciechowski for their editorial advice and constructive criticism; Kim Olin for her help in typing the manuscript; and Bill and Lucille Gray for their continued interest.

Paul A. LaViolette
Portland State University

Source Notes

The materials included in this book are slightly revised versions of previously published articles. They are reprinted here with the permission of the original publishers.

Chapter 1: "Man's Universe of Symbols"

"A Biologist Looks at Human Nature," *Scientific Monthly*, Vol. 82, no. 1, 1956, pp. 37–41.

Chapter 2: "Human Values in a Changing World"

"Human Values in a Changing World," in A. H. Maslow (ed.), *New Knowledge in Human Values*, New York: Harper, 1959, pp. 65–74.

"The World of Science and the World of Value." Address given at Central Washington State College, Ellensburg, Washington, in the spring of 1962. Published in *Teachers College Record*, Vol. 65, 1964, pp. 494–501.

Chapter 3: "Comments on Aggression"

"Comments on Aggression," *Bulletin of The Menninger Clinic*, Vol. 22, 1958, pp. 50–57.

"Human Values in a Changing World," in Maslow (ed.), *New Knowledge*.

Chapter 4: "Systems Perspectives on the Problem of Mental Illness"

"Some Biological Considerations of the Problem of Mental Illness," paper presented to the Institute for Schizophrenia held on October 1-3, 1958, in Osawatomie, Kansas. Published in *Bulletin of The Menninger Clinic*, Vol. 23, 1959, pp. 41-51.

Chapter 5: "A Definition of the Symbol"

"On the Definition of the Symbol," in J. R. Royce (ed.), *Psychology and the Symbol: An Interdisciplinary Symposium*, New York: Random House, 1965, pp. 1-10.

Chapter 6: "An Etymology of Symbolism"

"Reconsideration 1972: A Mini-History of the Concept of Symbolism," *Quarterly Bulletin of the Center for Theoretical Biology* (SUNY at Buffalo), 1972, pp. 153-161.

"On the Definition of the Symbol," in Royce (ed.), *Psychology and the Symbol*, pp. 10-16.

Chapter 7: "The Evolutionary Origins of Symbolism"

"On the Definition of the Symbol," in Royce (ed.), *Psychology and the Symbol*, pp. 17-30.

"The Mind-Body Problem: A New View," *Psychosomatic Medicine*, Vol. 24, 1964, p. 18. (Adapted from an address given at the Fifth Conference on Psychiatric Research, Harvard Medical School, Cambridge, Massachusetts, June 27-29, 1963.

Chapter 8: "The Mind-Body Problem: A New View"

"The Mind-Body Problem: A New View," *Psychosomatic Medicine*, Vol. 24, 1964, pp. 29-45.

Chapter 9: "General Theory of Systems: Application to Psychology"

"General Theory of Systems: Application to Psychology," *Social Science Information*, Vol. 6, December 1967, pp. 125–136.

"Theoretical Models in Biology and Psychology," *Journal of Personality*, Vol. 20, no. 1, 1951, pp. 32–33.

Chapter 10: "Toward a Generalized Theoretical Model for Psychology"

"Theoretical Models in Biology and Psychology," *Journal of Personality*, Vol. 20, no. 1, 1951, pp. 24–38.

Chapter 11: "Problems of Education in America"

"The World of Science and the World of Value," *Teachers College Record*, Vol. 65, 1964, pp. 501–507.

Chapter 12: "On Interdisciplinary Study"

"Democracy and Elite: The Educational Quest," a report written for the Committee on Interdisciplinary Education of the University of Alberta, Edmonton, Canada. Published in *Main Currents in Modern Thought*, Vol. 19, no. 2, November-December 1962, pp. 31–36.

Appendix

"A Biologist Looks at Human Nature," *Scientific Monthly*, Vol. 82, no. 1, pp. 35–37.

Introduction

Ludwig von Bertalanffy (1901-1972), originally a biologist, grasped the value of the organismic conception early in his career. He recognized that a system, whether it be an atom, a cell, a gestalt pattern, or an integrated universe of symbols, has holistic properties that are not found separately in its parts. Rather, these properties arise from the relations taken on by the parts in forming the whole. It was in the late 1920s that he emphasized under the title "organismic biology" the necessity of regarding the living organism as an "organized system" and defined the fundamental task of biology as the "discovery of the laws of biological systems at all levels of organization." This led him, in the 1930s and 1940s, to conceive of the notion of *general system theory*. This new *synthetic* approach to understanding nature he defined as "an interdisciplinary doctrine elaborating principles and models that apply to systems in general, irrespective of their particular kind, elements, and 'forces' involved" (in Laszlo, 1972, p. xvii).

Von Bertalanffy's aim was to achieve a general perspective, a coherent view of the "world as a great organization," a framework in which all disciplines could be understood in their place. However, such a world view would be incomplete if it did not provide a way of understanding and placing in perspective the most complex of systems: the human being. Consequently, in this book we find him mustering a wide range of fields, from linguistics and cultural anthropology to psychiatry and systems theory, in an attempt to synthesize a deeper understanding of human nature. His approach, which considers the world of symbols, values, and cultures as "being 'real' entities classifiable in the cosmic hierarchy of order," may succeed in bridging the age-old gap between the sciences and the humanities.

The essays that have been drawn together in this book were originally published in the 1950s and 1960s. Obviously, much has transpired since then in the fields of linguistics, psychology, and anthropology. However, von Bertalanffy was a revolutionary of sorts, one ahead of his times. Many of the concepts that he presented were heresies against the dominant dogma of the times; yet today these same ideas are rapidly becoming part of the humanistic movement that pervades all of the social sciences. We must acknowledge, therefore, that von Bertalanffy was a prophet of his times envisioning the form of things to come.

If we were to attempt to categorize these essays as falling under a particular field of study, probably it would be most appropriate to say that they belong to that broad field called epistemology. However, despite their generality, these studies may be conceived as falling into three main groupings: those centered around the theme of symbolism (Chapters 1-8), those concerned with psychology (Chapters 9 and 10), and those dealing with education (Chapters 11 and 12). Thus arranged, they fall into a logical rather than a chronological order.

Beginning in Chapter 1, von Bertalanffy addresses a question basic to anthropology: What is the prime factor that distinguishes human beings from other animals? He comes to the conclusion that, besides certain biological differences, the prime distinguishing characteristic of humans is their "creation of a universe of symbols in thought and language," that humans live "in a world not of things, but of symbols." Von Bertalanffy defines symbols as signs that are: (a) freely created, (b) representative of some content, and (c) transmitted by tradition. He proceeds to explain this definition and show how it allows one to distinguish symbolism, and human language in particular, from subhuman forms of behavior, such as the instinctual language of bees or the expressive song of birds.

When von Bertalanffy speaks of symbols, he connotes the mental phenomenon of symbolism; in particular, he refers to conscious representations such as thoughts and values. Neither Freudian nor Jungian symbols would be symbols by his definition. Because they arise from associative processes in the unconscious and lack consolidation, he suggests that Freudian symbols might best be termed "presymbols"; that is, they might be thought of as the raw material from which symbols arise.

We begin to see the brilliance of his approach when he extends his concept of symbolism to embrace the notion of higher level constructs: systems of symbols, or what he terms "symbolic universes." With this he is able to explain, among other things, why language, science, art, and other cultural forms are able to gain a relatively autonomous existence transcending the personalities and lifetimes of their individual creators.

This leads to the conception, developed in Chapter 2, of a human as a "denizen of two worlds," as a biological organism, but one that creates, uses, dominates, and is dominated by a higher world—the universe of symbols. This view, in a challenge to modern reductionism, states that human values cannot be derived from and ultimately reduced to biological values such as self-preservation and the sex drive.

Thus, Chapter 2 places the question of human values in a new perspective—one that views values as constituting the very fabric of the socio-culture. The loss of goals, or values worth fighting for, suggests von Bertalanffy, threatens the breakdown of established symbolic universes and contributes to the growing feeling of life's meaninglessness. Among other causes of the "sick society," he cites the current trend toward mass civilization, a society that does not recognize the value of the individual, but that for the sake of expediency utilizes techniques of mass manipulation.

Von Bertalanffy's view of a human as an *animal symbolicum* permeated many of his writings. For example, symbolism is a central theme in his essays on aggression (Chapter 3) and mental illness (Chapter 4). He suggested that acts of aggression, such as war and self-destruction, are not inherited from man's "animal nature" but arise within symbolic frameworks and, hence, are unique to humans. Moreover, mental illnesses such as schizophrenia, he felt, arise from disturbances in an individual's symbolic universe, a notion that calls into question the feasibility of searching for biochemical or physiological bases.

In Chapters 5, 6, and 7 we again confront the subject of symbolism, this time exploring the concept in greater depth. Chapter 5 deliberates further on what constitutes symbolism and discusses its various forms. Chapter 6 traces the historical development of the concept of symbolism and discusses the viewpoints of contributors such as Cassirer and Langer, whose works agree with von Bertalanffy's views on symbolism. Cassirer and Langer broached the subject from a philosophical angle, while von Bertalanffy independently asked the same questions from a biologist's point of view and arrived at quite similar results, only later to discover the affinity of his work with that of Cassirer and Langer. Such independent development of ideas from different starting points lends strong credibility to von Bertalanffy's arguments and ranks him as an important contributor to this field.

The analysis of symbolism is continued in Chapter 7, which grapples with the question: Why did symbolic systems evolve in the species *Homo sapiens*? What were the predisposing factors making this evolution possible? Von Bertalanffy believed that the development of symbolic activity by primeval man was a creative act and that empathy, hypostatization, and reification played a critical role. Tracing this evolution, he suggested that symbolism was originally employed at the pre-conscious or unconscious level and was part of early myth, magic, and ritual. He suggested that at this stage the ability to differentiate between the *I*, the animate *thou*, the inanimate *it*, and the symbol was not fully developed. The ego barrier, he believed, developed only gradually with the refinement of the representative function. Thus, self-awareness or individual consciousness in man would be a late product of culture, not unilaterally caused by, but standing in mutual interaction with, conceptualization and language.

This developmental approach led von Bertalanffy to the "perspectivist" view that mental categories represent certain perspectives of reality, yet are relativistic, being conditioned by both innate biological (species) factors and historical predispositions. Thus, the mythical categories of primitive man and the presently popular categories of Western thinking and science represent two of many possible ways of conceptualizing the world.

Von Bertalanffy effectively applied his general systems perspective in his attack on the classical mind-body problem, bringing it out of the realm of philosophical discourse and testing it against the knowledge of modern science (Chapter 8). He suggested that the Cartesian dualism between material things and conscious ego is not a primordial or elementary datum, but is a particular perception of reality based on a particular set of mental categories typical of seventeenth-century western Europe and resulting from a long evolution and development in the history of ideas.

However, von Bertalanffy noted that the Cartesian dualism still grips the fields of psychology and psychiatry where it masquerades, among other things, in the form of physicalistic models of brain function and as the dichotomy in psychiatric therapy between physical and psychological methods. To remedy this present state of confusion, he suggested that we relate the fields of psychology (mind) and neurophysiology (body) by postulating an isomorphism between their constructs and that we seek to unify these fields in theoretical principles that are generalized with respect to both.

Chapter 9 reviews a number of such constructs that include concepts such as: open system, differentiation, centralization, boundaries, and other organismic principles taken from the field of general system theory. Such principles portray man as an intrinsically active psychophysical organism possessing autonomous behavior. This view stands in contrast to the reactive conception portrayed by mechanistic models such as the classical stimulus-response scheme and the cybernetic feedback scheme.

Von Bertalanffy suggested certain new directions that should be taken to develop theoretical models in psychology (Chapter 10). Also, continuing his analysis of organismic principles from the previous chapter, he reviewed experiments on gestalt perception and on the functioning of the central nervous system and explained why computer, or cybernetic, models of brain functioning are grossly inadequate.

Chapters 11 and 12 are devoted mainly to the topic of education. Von Bertalanffy pointed out aspects of American psychology and educational practices that threaten to stifle individual creativity and scientific development and calls for a more interdisciplinary approach to education. He suggested that the world would make much better use of its resources if future scientists and engineers knew a bit more about the historical bases and sociological forces making up our civilization and if historians knew about the biological foundations of human behavior and the role of science in modern life.

In conclusion, I might add that if a holistic perspective of man, such as that which von Bertalanffy presented in the essays that compose this volume, were to be more fully adopted by our civilization, perhaps then the peaceful, humanistic world that we have long wished for could be realized sooner than we think.

Paul A. LaViolette

1
Man's Universe of Symbols

Everything in this world is nonsense; the whole of life is a plethora of ludicrous absurdities, one more fanciful than another. The crown is nothing, the ring is nothing, too. Each would mean nothing but nonsense and empty foolishness except to the eyes which behold the symbolism behind them. Yet they, because of their meaning, dominate the world. Only one form of metal there is, which is meaning in itself—the sword.
—Temple Thurston, 1909

What is unique in human behavior? The answer is unequivocal. The monopoly which man holds, which profoundly distinguishes him from other beings, is his ability to create a universe of symbols in thought and language. Except in the immediate satisfaction of biological needs, man lives in a world not of things but of symbols. A coin is a symbol for a certain amount of work done, or for the availability of a certain amount of food or other commodities; a document is a symbol of *res gestae* (things done); a book is a fantastic pile of accumulated symbols; and so forth ad infinitum.

To distinguish symbolism, and language in particular, from subhuman forms of behavior, the following definition might be proposed: Symbols are signs that are *freely created, represent some content*, and are *transmitted by tradition*. By "freely created" I mean that there is no biologically enforced connection between the sign and the thing connoted. In the case of conditioned reactions, the connection between the signal and the thing signaled is imposed from outside. It may be a natural connection, as when a child or the kitten has previously been burned. Or the connection may be arbitrary and imposed by the experimenter, as in the case of the Pavlovian dog that learned to secrete saliva when a bell was rung. In contradistinction, there is no biological connection between the word *father, pater, pere, otec* (or whatever the word may be in any language) and the person so designated. This does not imply that the choice of symbols is completely arbitrary; it is probably determined by psychological principles that are not well understood.

Furthermore, a symbol connotes or represents a certain content. This criterion distinguishes a symbol from language used in expressing a feeling. For example, a bird's song expresses and communicates to its mate a certain physiological and, we may be sure, psychological state, but it does not connote a thing. The barking of a dog warns of some danger, but it does not indictate whether the source is an intruding burglar or the neighbor's cat. Finally, symbolism and language are defined as being transmitted by learning and tradition. For example, the language of the bees, admirably described by von Frisch, is indeed representative. By means of intricate dances, the workers communicate to their colleagues the direction and distance of the place where food may be found. But this language is innate and instinctive. We can teach a dog all sorts of tricks, but we have never heard that a particularly clever dog has taught its puppies to do them.

Man's ability to use symbols was clearly made possible by the evolution of his forebrain (see Appendix). But, as to the origins of symbolism and human language, I have a strong suspicion that they might be found in imitation and verbal magic. An uttered sound may give some onomatopoetic image of an animal or person.[1] Consequently the sound will be identified with the original just as a puppet made from clay is identified with the enemy. Then, uttering the sound will govern the thing designated. For primitive man, an image, be it material or acoustical, is the same as the original and gives him control and dominance over it. This is the essence of sympathetic magic. The enemy can be killed if a needle is thrust into the clay image. In this way language may be born of magic, a process certainly infinitesimally slow in the beginnings, but man has had many hundred thousands of years at his disposal to come from an anthropoid to *Pithecanthropus, Sinanthropus,* and *Homo sapiens.*

Whatever the origin of symbolism, its consequences are enormous. First of all phylogenetic evolution, based on hereditary changes, becomes supplanted by history, based on the tradition of symbols. Normally in the biological sphere, progress is possible only at a rate determined by the slow pace of evolution. For example, ant societies have remained unchanged for the past 50 million years. In contrast, human history has a time-scale for change on the order of generations. In fact it may even be thought that the scale of cultural time is logarithmic, rather than arithmetic, with changes taking place at an ever-increasing pace.

A second consequence is that corporeal trial and error, as found in subhuman nature, becomes replaced by reasoning—that is, trial and error in terms of conceptual symbols. An animal placed in a maze, faced with a complicated lock, or confronted with some other problem, runs around until it finds the way out. It tries until the solution of the problem is discovered by chance. Man, in a corresponding situation, sits down and thinks. That is, he experiments, not with the things themselves, but with the symbolic images of the things. He scans different possibilities, accepts the apparently successful solution and discounts those that seem ineffective, without laboring to try them materially.

A third and even more profound consequence of symbolism is that it makes true purposiveness possible. This true, or Aristotelian, purposiveness is unique to human behavior and is based on the fact that the future goal is anticipated in thought and determines actual behavior. Of course, "purposiveness" in a metaphorical sense, that is, regulation of function in the way of establishment and maintenance of organismic order, is a general characteristic of life. It is based on such principles as equifinality of the steady state, homeostatic feedback, learning by trial and error and by conditioned reflex, evolutionary selection, and so forth. But even in the most amazing phenomena of regulation and instinct we have no justification for the assumption that these actions are carried through with foresight of the goal.

The Magic of the Algorithm

So long as symbols stand alone they are unproductive and do not convey more information than what is contained in the individual symbols. Thus, in the flag-language used by seafarers each flag symbolizes a certain fact or command. An array of flags is just the sum of the individual meanings. This is profoundly altered if symbols are combined and related according to established rules of a "game." Then, the system of symbols becomes productive and fertile. With a suitable choice of terms and of rules presupposed, a "grammar," we can handle the symbols, the "vocabulary," as if they were the things they represent. If the symbols, as well as their grammar, are well chosen, the result of the mental operation of symbols will correspond to that of the real course of events. The consequences of the images will be the images of the consequences, to use Heinrich Hertz's expression. In this way a true magic is possible with systems of symbols. We can predict facts and relationships still unknown, can control still unrealized combinations and natural forces and so on.

A system of symbols related according to preestablished rules may be called an algorithm. The simplest example of an algorithm is the mathematical system of decimal notation, popularized by a man named al-Khowarizmi. In Roman numbers, even a trivial multiplication, such as LXXVI × XCIII, is quite a formidable operation. However, the simple trick of regarding the last digit of a number, in Arabic notation, as meaning units, the second as meaning tens, and so on, and of writing the corresponding figures in columns, makes the operation child's play. Thus, an algorithm is essentially a "thinking" machine, a means of performing operations on symbols that give results difficult or impossible to attain otherwise. Calculating and thinking machines, mechanical or electronic, are the materialization of algorithms. The symbolic system of language, and particularly of the artificial languages called mathematics and science, develops into a colossal thinking machine. An operational command (an hypothesis) is fed in, the machine starts to run, and eventually, by virtue of preestablished rules relating the symbols, a solution drops out, a product which was unforeseeable to the individual mind with its limited capacity. This is the general character of scientific reasoning, be it a simple arithmetic operation or the solution of a differential equation, the prediction of still undiscovered planets and chemical elements, or the construction of some masterpiece of modern technology.

The symbolic universe becomes, so to speak, more clever than man, its creator. It wins an autonomous life of its own, as it were. The development of the Roman law, the British Empire, the atomic theory from Democritus to Heisenberg, or of music from Palestrina to Wagner, is certainly borne by a number of human individuals. But it shows an inherent logic that widely transcends the personalities of its creators.

However, besides these triumphs of symbolism, there are also its pitfalls. The conceptual anticipation of future events, which, on the one hand, allows for true purposiveness, is at the same time the origin of anxiety, the source of man's fear of death, which is unknown to brutes. The invention of the symbolic world was the Fall of Man. The notions of sin and evil arise with the invention of symbolic labels attached to certain forms of behavior. Moreover, if there is a clash between the symbolic world built up as moral values and social conventions, on the one hand, with basic biological drives, on the other, then the situation of neurosis arises. Somewhat extending the narrower definition of Freud, it seems that a neurotic situation results from the conflict of a symbolic universe with biological drives, or of opposing symbolic worlds.

As a social force, the universe of symbols, which is unique in man, creates the sanguinary course of history. Thus man has to pay for the uniqueness that distinguishes him from other beings. The tree of knowledge is the tree of death. War is a human invention, not a biological phenomenon. It is not the continuation of the omnipresent biological struggle for existence. Even if nature were "red in tooth and claw," which it is only to a limited extent, organized intraspecific warfare for the most part would be unknown in the subhuman world. Apart from the rather rational strifes of savages who go out and kill enemies in order to eat them, war is caused by illusions of grandeur, ideologies, economic factors based upon symbol-charged values, and religion. This, however, leads to the ingratiating conclusion that war is not a biological necessity and that it would be avoidable if mankind were to put its symbolic faculties to better use.

Revolt of the Masses

The unique characteristic of human behavior is the ability to make decisions at the symbolic level. This, of course, does not mean that conditioned behavior is negligible. Any human achievement, from toilet training to speech, driving a car, or learning calculus and theoretical physics, is based on conditioning. Nevertheless, the specificity of man rests on rational behavior—that is, behavior directed by symbolic anticipation of a goal. In modern man, however, this *vis a fronte*, to use Aristotle's terms, consisting of goals which the individual or the society sets itself, is largely replaced by the primitive *vis a tergo* of conditional reaction. The basic symptom of present society seems to be the "uprising of the masses," or put another way, the "return of the conditioned reflex."

The modern methods of propaganda, from the advertisement of a toothpaste to that of political programs and political systems, do not appeal to rationality in man but rather force upon him certain ways of behavior by means of a continuous repetition of stimuli coupled with emotional rewards or punishments. This method is essentially the same as that applied to Pavlovian dogs when they were drilled to respond to a meaningless stimulus with reactions prescribed by the experimenter. Not that this method is new in human history. What is new, however, is that it is applied scientifically and consistently and so has an unprecedented power. The modern media of mass communication, newspapers, radio, television, and so on, are able to establish this psychological constraint almost without interruption in time and reaching all individuals in space with maximum efficiency. If a slogan, however insipid, is repeated a sufficient number of times and is emotionally coupled with the promise of a reward or the menace of punishment, it is nearly unavoidable that the human animal establishes the conditioned reaction as desired. Furthermore, to apply this method successfully, the conditioning process must be adjusted to the greatest common denominator; that is, the appeal has to be made to the lowest level of intelligence. As a result, individual discrimination and decision become replaced by universal conditioned reflexes. What emerges is mass-man. *Brave New World* and *1984* are but paraphrases of this theme.

It seems that behavioral science is supposed to contribute to the pressing problems of our epoch. However, American psychology, in its various forms from behaviorism to recent pseudo-"humanistic" developments, has consistently denied what is specifically human in man's psychology and behavior. It has given no answer to "human" problems and, in effect, has contributed to the bestialization of our age of non-culture. Beside the menace of physical technology, the dangers of psychological technology are often overlooked; perhaps even more dangerous than the material existence of the bombs are the psychological forces that may lead to dropping them. As we try to put atomic energy to peaceful use, it may even be more urgent to put to intelligent use the psychological mechanisms revealed by behavioral science.

2
Human Values in a Changing World

What can we say has become commonly accepted as human values? This very question is disquietingly symptomatic; asking the question implies that values have become doubtful and are not taken for granted anymore. Hence, I propose some sentences from Nietzsche's *Will to Power* as a suitable starting point for this discussion: "What I relate," Nietzsche wrote in 1868, "is the history of the next two centuries. I describe what is coming, what can no longer come differently: the advent of nihilism.... Our whole European culture is moving for some time now, with a tortured tension headlong.... Why has the advent of nihilism become necessary? Because the values we had hitherto thus draw their final consequence; because we must experience nihilism before we can find out what value these 'values' really had."

Nietzsche regarded Christianity as the value system to be discarded and replaced by a new one. One possible answer to Nietzsche's demand for new values was the notion of progress, the belief that science and technology will carry mankind into a paradisiac future. It can be stated with no elaborate discussion that, up to comparatively recent times, the belief in progress was the dominant ideology of our civilization, and that now we have become doubtful. As this question unavoidably leads to platitudes, let us quickly dispose of it. It is an obvious fact that now I can fly from Los Angeles to Boston in eight hours, whereas twenty years ago this trip would have taken four days by train; that the number of automobiles, washing machines, and television sets is multiplying; that the situation of the American worker today is not comparable to that of fifty years ago, or even to that of his European colleague; that the average life span has increased some twenty years in the past few decades, and so forth ad infinitum.

Let us also pass over the dangers inherent in this development: the hydrogen bomb; the possible social and psychological implications of automation; the Malthusian menace of further overcrowding our planet in consequence of the advances of modern medicine; and all the rest. Whether the sum total of human happiness and misery has increased, decreased, or remained approximately constant during the course of human history nobody can tell because there is no yardstick to measure it. Considering the two total wars and the number of minor ones within the life span of one generation, the balance does not automatically jump in favor of our epoch as against the Thirty Years War, the Spanish Inquisition, or the French Revolution.

Life and history are no idyll. When we look back, we find the greatest geniuses, from the prophets in the Bible to Sophocles, Dante, Michelangelo, and even Goethe, filled with dismay about their times; with what modern philosophers would call existential anxiety, deep-rooted doubts about the meaning and goals of life. Nevertheless, we hardly commit an exaggeration when saying that there never was a more profound insecurity about our directions than in present times; a deeper, more all-pervading gap between facts and values, between the world which is and the world which ought to be.

The early Christians in the Roman catacombs did not know whether they would see the next day, but they did know that their martyrdom granted the Crown of Life. The Italian Renaissance was politically one of the most atrocious episodes in history, but it sublimated its gore and cruelty into Giotto's frescos at St. Francis's in Assisi and into the Sistine Chapel and the triumphal glory of St. Peter's. The French Revolution slaughtered thousands at the altar of Liberty, but it brought a new idea into the world which will not perish. We, with all our skyscrapers, space vehicles, comfortable homes, economic abundance, our cars, and doubled life span, are not so fortunate. Whether the abyss of atomic annihilation will devour us or whether we manage precariously to dance at its brink, if everything is said, our creed is that of Iago in Verdi's *Otello—Sento il fango originario in me; e poi? La morte e nulla* (I come from primeval slime, and my destiny is death and nothing).

Let us not believe that spiritual questions are superannuated in an age of technology. There is an old saying that God is with the stronger battalions. Modern inventiveness has gone only so far as to replace infantry battalions with atomic bombs. In the last resort, however, it is always a system of values, of ideas, of ideologies that is decisive. It was an idea that founded the United States, even though little bands of settlers were fighting a mighty empire. An idea was victorious in the group of poor subversives called Early Christians because their imperial adversary governed the world but had lost his basic concept of existence. Napoleon's soldiers, hungry and in rags, conquered Italy and Europe with ideas and determination. We have no reason to assume that this law of history has changed. What has lost its historic meaning will not survive. Military hardware, including the most advanced superbombs, will not save us when the will to live, the guiding ideas or values of life, have subsided. This is one of the few safe conclusions from history.

Poignant Paradoxes

I would say that one excellent means to grasp the spirit of the time is by reading newspapers and magazines. On one single page, we may observe all the striking contrasts besetting our time. The report of the latest space exploit is sandwiched between the latest murder and Hollywood divorce, between Kennedy and Khrushchev, or between the new car model and nuclear annihilation. Contemplate, for a moment, the symbolic meaning of such arrangements.

The conquest of space is not only a most brilliant achievement of science and technology; it is the fulfillment of a millenial longing of humanity, first expressed in the myth of Daedalus, and visionarily anticipated by utopianists from Leonardo da Vinci to Goya and Jules Verne. Oswald Spengler, the philosopher and author of the *Decline of the West*, states that space is the *ur-symbol,* the deepest and most decisive symbol of the occidental mind, expressed in all manifestations of culture—the longing for space in painting and in music, space-conquering diplomacy, voyages, physical theories, and innumerable others. Eventually and after centuries, this *ur-symbol* materializes in its definitive, never-before-believed form and is submerged by trivia, by appeals to what is lowest in human nature, by the cheap sensations provided by an almost subhuman killer and the ephemeral amours of a doll.

Ours is the affluent society, so we read, and we have the highest standard of living ever achieved. We are bombarded with astronomical figures of gross national product—$20 billion for the first trip to the moon, $11 billion for packaging wares to make them appetizing to the buyer. But we also read of $100 billion which would be required but are not available for slum clearing; we read that 57 per cent of people over age 65 live on less than $1,000 per year in cold-water flats; that 10 per cent of Americans are functional illiterates.

What is perhaps the most remarkable symptom: Economic opulence goes hand in hand with a continuous increase in the rate of crime, especially juvenile delinquency. And the psychotherapists tell us that, besides the classical neuroses (caused by stress, tensions, and psychological trauma) a new type of mental sickness has developed for which they have even had to coin a new term—*existential neurosis,* that is, mental illness arising from the meaninglessness of life, caused by the lack of goals and the lack of hopes in a mechanized mass society.

Society as Patient

The psychiatrist is wont to speak of split personality as a classical symptom of mental disease. If anything, our society is a split personality, not simply a fine Dr. Jekyll and hideous Mr. Hyde, but split into an enormity of disorganized and antagonistic parts. This is the reason why analyses of modern society are no longer merely book titles, but have become part of everyday language: from the *Decline of the West* to *Brave New World, 1984, Organization Man, Hidden Persuaders, Waste Makers, Status Seekers,* and many others. This literature in itself is a symptom or symbol; nothing comparable has existed in history, except perhaps in the analogous time of the decay of the Roman Empire. Like physicians examining an individual patient, these modern diagnosticians of society observe different symptoms, use different tools and terms, sometimes err or exaggerate their findings. On the whole, however, their analyses are like a battery of laboratory tests, adding up to a consistent picture. It may be expressed in one brief sentence: We have conquered the world, but somewhere on the way, we seem to have lost our soul.

In more realistic terms, this means that we have lost, or lost sight of, those guiding lights for the formation of our lives which are called human values. Unfortunately, the theory of values is one of the most difficult, obscure, and controversial fields in philosophy and behavioral science. The best we can do is start with an operational definition of value and see how far we get. That is, we adapt the definition to our purposes, keeping in mind that, for this very reason, it will not be uncontradictable.

A Theory of Value

So let us posit: Values are things or acts which are chosen by and are desirable to an individual or to society within a certain frame of reference. Although it is admittedly tentative, every word matters in this definition. We obviously have to include both objects and acts. Material things like dollar bills or Picasso paintings and immaterial qualities like the goodness of a charitable act are obviously values. We further have to introduce the element of choice. Where there is no choice, there is only necessity, not value. In somewhat different terms, whatever is taken for granted neither is nor has a value. For example, to a perfectly healthy person or animal, health is not a value but is simply taken care of by biological functions. Only if we envisage possible danger and can do something about it does health become a value. Prolongation of human life is possible; physical immortality is not. The first is a value, not the second. We have to pay for food, not for breathing air; therefore, the first has value, the second not, even though it is equally indispensable for life. To the Aztecs, because gold was abundantly available, it had no particular value. To the Spaniards, it was eminently desirable, so they liquidated Montezuma and his Indians. A postage stamp is worth a few pennies to the letter writer; it may be worth thousands of dollars to the collector because it is particularly desirable within the framework of philately.

Our criteria equally apply to actions. In the course of a day, we perform innumerable actions which have no value involvement at all. Only where there is both choice and preferability do value judgments appear. Nobody cares what way I choose to arrive at my office—at least so long as I do not commit an undesirable act like a traffic violation; but my consistently coming late or early may be evaluated. On the other hand, if I fall into the river and save myself by swimming, this is not considered to be a moral act because it is presupposed that I had no choice. If another person falls into the river and I save him, I may get a medal because it is presupposed that I did have a choice and took the socially desirable action. And so forth ad infinitum.

From this infinite array of preferences and evaluations, mankind, starting in some prehistoric stage and continuing to the present day, singled out some very general and abstract notions which became values *par excellence*. Pleasure, social virtues, goodness, truth, beauty, deity are a few of them. It is the objective of a theory of value to elucidate where they came from, what they mean, from what ultimate concept they can be derived, and what are their consequences for human behavior and society.

Biological and Human Values

In discussions of human values the following question is often posed to the biologist: Can human values be derived from and reduced to biological values? It may be surprising, or even shocking, to some readers that the answer of at least one biologist is a well-considered "No."

The basis of the parallel between organic evolution and human progress is what is known as the postulate of reductionism: that is, that biology should eventually be reduced to physics and chemistry and, correspondingly, the behavioral and social sciences to biology.[1] Put in a somewhat different way: A living organism, it is assumed, is an intricate physiochemical system; hence, human behavior is a particularly involved complex of the ways and factors of behavior present in subhuman species. Human values are viewed as being derived from and ultimately reducible to biological values which essentially entail maintenance of the individual, survival of the group, and evolution of the species.

This basic doctrine can and has been formulated in many different ways. For example, it is the classic philosophical doctrine of hedonism, maintaining that pleasure is the ultimate good. It is also Freud's doctrine that behavior is governed by the pleasure principle and the principle of sustaining the homeostatic equilibrium of the organism in answer to changing environmental influences. Generalizing the original physiological meaning of homeostasis, the terms of psychological and sociological homeostasis were introduced: that is, the ultimate goal of behavior is to maintain the psychophysical organism in a biological, psychological, and social equilibrium. Still other terms for the same idea are psychological and social adjustment or adaptation; from here originate the philosophy of conformity and the ideology of so-called progressive education, both proclaiming social adjustment or equilibrium with existing society as the ultimate goal.

Regarding this theory, I am in fundamental disagreement, not because of theological or metaphysical prejudices, but because the theory just does not fit the facts. Human behavior is not simply directed toward the release of tensions. Boredom, emptiness, and *taedium vitae* may also be important psychopathogenic factors. Also, a large part of behavior, such as play and exploratory activities, creativity, and culture in general, simply doesn't fit in the scheme. Human beings (and organisms in general) are not stimulus-response machines, as the theory presupposes; immanent activity, along with so-called function pleasure, is an important part of behavior. Life and behavior are not simply utilitarian, trying to come to a so-called equilibrium with minimum expense of physical and psychic energy. This is not even true of organic evolution, which often produces fantastic formations, behavior patterns, colors, and whatnot, far exceeding mere survival and economic principles of adaptation. It is even less true of man, where not by the wildest flight of fancy can the creativity of an artist, musician, or scientist be reduced to psychological and social adjustment, nor can the self-sacrifice of a martyr be reduced to the principle of utility. The whole of human culture—whether Greek tragedy, Renaissance art, or German music—simply has nothing to do with biological values of maintenance, survival, adjustment, or homeostasis.

To take a somewhat different approach, I believe that the field of science, speaking in the way of gross oversimplification, consists of three major levels: physical nature; organisms; and human behavior (individual and social). In the way of an idealized graph, the scale of nature may be represented as is shown in Figure 1. I shall not discuss the question of how these levels are connected. It appears that they represent distinctive steps, although not discontinuities, somewhat in the way the mathematician speaks of step functions. There are intermediates between the levels of inanimate and living nature, such as the viruses, and again intermediates between animal behavior and the symbolic behavior characteristic of man.

It is well accepted that the fields of physics and chemistry form an indispensable groundwork for biology and that, in a

Figure 1 Levels in nature

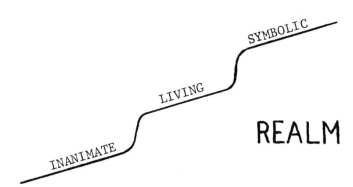

similar way, biology forms an indispensable groundwork for the science of human behavior. Nevertheless, the notion of emergence is essentially correct: *Each higher level presents new features that surpass those of the lower levels.*

Man in Two Worlds

Man, as the old saying goes, is a denizen of two worlds. He is a biological organism with the physical equipment, drives, instincts, and limitations of his species. At the same time, he creates, uses, dominates, and is dominated by a higher world which, without theological and philosophical implications and in behavioral terms, can best be defined as the universe (or universes) of symbols. This is what we call human culture; and values—esthetic, scientific, ethical, religious—are all part of this symbolic universe. This is what man tries to achieve beyond satisfaction of his biological needs and drives; in turn, it governs and controls his behavior.

The fact that man lives in a universe, not of things, but of symbolic stand-ins for things, indicates the difference between biological "values" and specifically human values. Compared to the biological traits or "values" which may be categorized as either "useful" or "harmful" for the survival of the individual and the species, what we call human values are essentially symbolic universes that have developed over the course of history. This conceptualization is applicable to any field of human activity, be it science, technology, art, morals, or religion. These symbolic universes may be adaptive and utilitarian in the biological sense, as when technology allows man to control nature. They may be indifferent, such as Greek sculpture or Renaissance painting, which hardly can be claimed to have contributed toward better adaptation and survival. On the other hand, they may be outright deleterious if the breakdown of an individual's "little" symbolic universe leads him to commit suicide, or if the conflict of larger symbolic worlds leads to war and extermination on a large scale.

The man-created symbolic universe partly depends on categories which are universally human, and partly on categories developed historically within a certain civilization

(Bertalanffy, 1955). The first fact accounts for the conformity of a Golden Rule of behavior common to all higher religions; the second, for the solipsisms of moral codes. Moral values are indeed different for modern Europeans as compared with Pueblo Indians (Benedict, 1934); so much so that normal ways of behavior in one culture would be considered schizophrenic in another one. This all depends on the symbolic frameworks of the respective cultures or even on different frames of reference within a given culture; what is penalized as murder in civilian life is labeled as heroism in the frame of reference of war.

Nihilism—The Breakdown of a Symbolic Universe

We spoke earlier of existential neurosis—mental sickness resulting from the meaninglessness and emptiness of life and a lack of desirable goals. Why has life become devoid of meaning and goals at a time of affluence and high standards of living, whereas it apparently had meaning and goals in times incomparably poorer in their economic and technical resources? The best answer I am able to find is that this phenomenon, termed *nihilism* by Nietzsche, is nothing else than the imminent breakdown of a symbolic universe. You will find this definition applicable to any field of human behavior and activity. The economic symbol of money has lost its connection with reality; a banknote does not represent any more a fixed quantity of gold or of commodities but is subject to continuous re-evaluation, to inflation, sometimes at an astronomical scale. Art, music, even education and science tend to lose their intrinsic value and to retain only their utilitarian or snob value. Art used to be a symbol-system representative of a certain period in a certain culture. Today's "art" seems to extend from the finger-painting of a chimpanzee, reproduced in *Life* magazine and presenting a good example of modern non-representative pictures, to the homey covers of the *Saturday Evening Post*.

Even the symbolic universe of science, which is about the only solid thing we have, is shaky in certain aspects and places. The same symbol, democracy, means exactly the opposite when uttered in the West or in the communist world. The symbolic system of religion which, abstracting from its intrinsic values, at any rate has developed organically in the long course of history, is supplanted by kaleidoscopically changing pseudo-religions, be they scientific progress, psychoanalysis, nationalism, soap opera, or tranquilizers. The need for a value system of religion or at least a secular ideology remains unfilled; the substitute or *ersatz* is Christianity serving as a social affair or status symbol, advertised with all the tricks of the trade.

All this does not mean that we necessarily behave worse than our predecessors, or that our predecessors were better than we are. It does mean, however, that there used to be established symbolic standards which were taken for granted even by the trespasser, sinner, and reformer, while now they seem to be disappearing.

I will illustrate this only by a few examples taken at random. Take, for instance, a problem in which we educators are immediately concerned, that "Johnny can't read" and, what is worse, that Johnny doesn't much care to read. In spite of general and supposedly progressive education, we feel the advent of a new illiteracy, nourished by comics, television, and talkies. At the adult level, it is the same phenomenon when the so-called intelligentsia is labeled as a bunch of sissies or Communists, and Babbitt becomes the image of ideal humanity. What else is this than the breakdown of a symbolic universe, laboriously and under a thousand pains erected in the course of history?

Another facet is what I have called the return of the conditioned reflex (see Chapter 1). Still another aspect is what might be called cultural regression, using the term precisely in the psychoanalytic sense. Partly, even the term "mental fetalization," extending Bolk's use of the word, would seem in place. The comic strip, the peeping Tom in the modern form of scandal magazines, the touching infantility of television, the well-filled refrigerator as a sort of nourishing womb, the penis symbolism of the Cadillac,[2] the father image of Eisenhower—all these and many other things are nothing but regression to infantile states. This does not mean that automobiles, refrigerators, or Eisenhower may not in themselves be excellent things; but it does mean that the attitude is not that of what psychoanalysts call a mature ego. No wonder, then, that the breakdown of the symbolic universe leads to the experience of being lost in a meaningless world. This experience, as formulated by individuals of a high intellectual standard, is existentialist philosophy.

For the many who lack sophistication, there are two other outlets: crime and mental disease. Juvenile delinquency reaching a peak under optimum economic conditions—what else can it mean than the practice of a philosophy of meaninglessness? And there is that other outlet—mental ailment to the tune of some 10 million cases in the United States, which comprise 52 per cent of all hospitalized patients (Menninger, 1957). It is often maintained that the stress of modern life is at the basis of the increase of mental derangement. This stress is bad enough, but the theory is demonstrably false. It is a statistical fact that in times of extreme stress, as, for example, the blitz in England, neurotic disorders decrease rather than increase (Opler, 1956). Measured in terms of stress, all of Europe, after war and postwar experience, should be a gigantic madhouse, which patently it is not. It appears that the shoe is rather on the other foot: Neurosis reaches its peak not when biological survival, but rather, when the symbolic superstructure is at stake. When life becomes intolerably dull, void, and meaningless, what can a person do but develop a neurosis? These neurosogenic or even psychosogenic conditions can even be reproduced in the laboratory, as in the well-known Montreal experiments (Bindra, 1957) where subjects, isolated from incoming stimuli, developed a model psychosis with intolerable anxiety, hallucinations, and the rest.

It is well understood that man is a creature seeking satisfaction of his biological needs (i.e., food, shelter, sex, an amount of security for his biological and social existence); however, he also lives in the higher realm of culture which is defined by the very fact that it transcends biological needs. Tradition, status in society, full realization of potentialities, religion, art, science —these are a few of the needs deriving from man's cultural existence. Starvation at this symbolic level leads to disturbances of the physical organism. This is a well-established fact of psychopathology.

The diagnosis of the sick society, then, is quite simply that it provides more or less abundantly for the biological needs but starves the spiritual ones. All the catchwords I have mentioned—from the Uprising of the Masses to Status Seekers to the Organization Man and so forth—are variations of this one theme. In modern mass civilization, tradition, which made the hard life of a peasant in the Alps tolerable and even enjoyable, is replaced by the titillation of ever new sensations, the cruder the better; from oversexed movies to television bloodies, from the hope of fatal events in the boxing ring to dangerous play (even possibly with atomic warfare) serving as distractions from a dull life. Status-seeking is a perfectly normal human ambition, having its precursor even in gregarious animals. But as there is no real status in mass civilization (which does not recognize the value of the individual) this yearning can only be satisfied by empty and often silly status symbols—the amount of chromium on the annually traded-in car, higher living than the Joneses, and a larger swimming pool. The human faculty of rational decision is replaced by biologistic factors, by conditioning like that of laboratory dogs and rats, and by exploitation of the unconscious—all brought to mastery in advertising, by hidden persuaders, motivation research, and human engineering in general. The system works to the profit of business because it is psychologically easier to be pushed by conditioned reflexes than to act with reason. But the loss of psychological freedom is paid for by a loss of goals worth fighting for and, consequently, by a feeling of emptiness and meaninglessness.

Is There a Remedy?

Such considerations, which could be continued indefinitely, are apt to show that culture, i.e., the framework of symbolic values, is not a mere plaything for the human animal or a luxury of the intelligentsia; it is the very backbone of society and, among many other things, an important psychohygienic factor.

What, then, can we offer for a re-evaluation of human values? For a minority, there is what Maslow has so aptly labeled and described under the term "peak experience" (Maslow, 1959). This, however, is reserved for an esoteric few. I am neither concerned with, nor competent on, theological matters. However, looking at things from a detached, scientific and historical viewpoint, it appears that the value system of Christianity has proved to be singularly persistent and adaptable to varying cultural frameworks. Of course, this means that I am in contradiction to Nietzsche, with whom I started these considerations.

In summary, the diagnosis I have tried to make may appear pessimistic. There are, however, some hopeful aspects. Present developments clearly show that these problems are no longer by-passed and neglected but are a matter of serious concern. Where there is insight and a will, there might be a way. A new symbolic universe of values must be found or an old one reinstated if mankind is to be saved from the pit of meaninglessness, suicide, and atomic fire.

3
Comments on Aggression

It appears that, among psychoanalysts, there are wide differences of opinion regarding the problem of aggression. I am not a psychoanalyst and do not belong to any particular psychoanalytic school. If my contribution is to be of any value, I believe it is not in taking sides, but rather in contributing material and viewpoints from my own field and lines of thought. The first question, which is partly biological and partly semantic, appears to me to be: What is meant when we speak of libido and aggression, life and death instinct, Eros and Thanatos?

When we speak of a "life" or "death instinct," we actually mean an ensemble or sum total of tendencies which, in commonplace terms, are in the line of love or in the line of hate. Similarly, animal instinct, such as nest building or mating, are actually composites of many mechanisms and actions—innate releasing mechanisms, as Lorenz (1943) calls them. Such lumping together under a common heading of many and diverse trends and traits is quite legitimate scientifically. We may compare it with the procedure of physiology when innumerable processes of building up and breaking down in the organism are lumped together under terms like anabolism and catabolism. However, we must be aware of what we do. We must not hypostatize these collective ideas or nouns into metaphysical entities, into a mythology or demonology, as if human beings were governed by metaphysical powers called life and death instincts.

It is difficult to draw a borderline between destruction and construction, aggression and productive work. To a large extent, destructive tendencies, or "instincts" are perfectly normal and are necessary for the survival of the human individual and species. They only have to be channeled into lines that are healthy for the individual and acceptable to society. Hence, as Karl Menninger (1942) has aptly shown, a destructive element is present in almost all our activities, from tilling the soil and raising and slaughtering animals for food to Michelangelo's violent attack on marble—everywhere destructive energies, some of them employed for constructive goals.

Unfortunately, we do not live in a world where the maxim *Love thy neighbor as thyself* is practicable. Ours is a world governed by the struggle for existence. The real problem enters if we see, in the human species, destructive tendencies and behavior that apparently are not in the service of the preservation of the individual or the species: Such behavior including self-destruction is termed "essential aggression," and according to Waelder (1956), cannot be explained without postulating a destructive drive, or a "death instinct" (according to my own terminology).

Destructive tendencies in the human species and in human civilization may be fostered by what in biology is called domestication. Conditions of domestication in the broad sense, that is, conditions where a species is relatively sheltered, allow for nonadaptive trends and in particular for intraspecific competition which, under the severe pressure of selection in wild life, would not be possible. For example, it is a well-known theory that over-ornamentation in animals, like the heavy antlers of deer and the magnificent plumage of birds of paradise, has arisen through sexual selection. Presumably, under relatively sheltered conditions there might be an intraspecific competition of the females for the most handsome males, eventually breeding colors, formations, and the like, the value of which for the survival of the species is doubtful or even definitely negative. Thus, intraspecific competition and aggression may be possible under conditions of domestication, while in a species standing in heavy competition with others it would soon lead to extinction. On the other hand, Lorenz (1943) has emphasized that gregarious animals of prey, like wolves, which hunt in packs, appear to have inbuilt instincts of chivalry. The defeated, by certain behavior, shows his surrender and is then pardoned by his opponent. Such behavior obviously is in the interest of the survival of the species. So domestication may foster intraspecific aggression which, in wild life and under severe interspecific competition, would soon be eliminated by selection.

Now I wish to offer a thesis which, so far as I can see, transcends the orthodox psychoanalytic view. I state it thus: There is no doubt about the presence of aggressive and destructive tendencies in the human psyche which are of the nature of biological drives. However, the most pernicious phenomena of aggression, capable of transcending self-preservation and self-defense, are based upon a characteristic feature of man above the biological level, namely, his capability of creating symbolic universes in thought, language, and behavior. In a somewhat different formulation we might say: Only a minor part of so-called essential destructiveness, such as crimes of violence, the destructive mob, the self-mutilation of the maniac, is purely on the level of the primary process. The much more devastating part is essentially connected with secondary processes.

If I speak here of symbols, I mean something different from the use of the term "symbol" in psychoanalysis, e.g., phallic symbols, or the symbolism of dreams. I mean rather a phenomenon which is obviously a unique characteristic of man, distinguishing him from other living beings. As was mentioned earlier (see Chapter 1), apart from the gratification of biological needs which man shares with other animals, he lives in a world of symbols. A word or a concept is a symbol for a thing or a relationship. Language or a book is a fantastic compilation of accumulated symbols. A new Cadillac is to a large extent not a vehicle for faster transportation than the old family Ford, but a symbol of social status. A social or moral code is a symbolic system of rules of behavior. Art, science, religion, political programs are all systems of symbols linked together by suitable rules of the game.

However, one characteristic of the symbolic universes created by man should be mentioned in this context. The symbolic worlds of social conventions, of morals, of religion, of art or of science, transcend the individual psychology of the human beings who have created them. They win, so to speak, an autonomous life of their own (see Chapter 1). The development of Roman law, of the British Empire, of the theory of atoms, or of occidental music were certainly due to the efforts of a number of human individuals. But they have laws of development and an immanent logic of their own that transcend the "petty" personalities of their creators. The same applies to customs, social institutions, moral conventions, political systems, and the rest.

This symbolic world, characteristic of and unique to man, is partly covered by Freud's concept of the superego, but not all of its implications are fully envisaged. One particular consequence is the predominance in human behavior of what may be termed *quasi needs*. Human behavior, to a large extent, is not governed by the primary and basic biological needs granting the survival of the individual and of the species, that is, the need to gratify hunger in order to keep the organism going, and the sexual drive in order to maintain the species. Rather, human behavior is governed by quasi needs, that is, needs arising within a certain symbolic framework. These quasi needs extend from "keeping up with the Joneses," the problem of the new car, or the cocktail party which must be given, to world-shaking political programs and the problems of the genius and martyr in science, art, and religion. I believe that, in particular, the problem of aggression cannot be satisfactorily visualized without taking into account this aspect of human behavior. Indeed, most of the examples quoted by Waelder (1956) for "essential destructiveness" or for a "primary aggressive drive" fall precisely in the symbolic level.

There is a great variety of aggressive phenomena. At the one extreme of the spectrum one may classify the violent outbursts in psychosis caused by biochemical disturbances, for example, the Viking *Berserksgang* (going *berserk*, or running amok). Rosen (1956) offers this as an example of such essential or primary aggression. However, according to a recent investigation, it appears probable that the Viking's going berserk falls into the category of drug-produced psychosis. Fabing (1956) has recently shown that it probably was caused by eating a poisonous mushroom of the genus *Amanita*, containing a substance which, like lysergic acid, produces a psychosis-like state. Phenomena of this sort then obviously are not an expression of an innate primary aggression or destructive drive, but rather a pharmacological effect, obtained within a certain ritual. At the other extreme stand such examples of "essential destructiveness," quoted by Waelder (1956), as Hitler or Stalin. It is to be expected that there are psychoanalytic roots to the destructiveness of such dictators stemming from their early childhood experience, such as unsatisfactory mother-child relationships. However, under ordinary circumstances such psychoanalytic factors would have produced unpleasant contemporaries, neurotics, or, at the worst, criminals. If Hitler or Stalin, driven by "essential destructiveness," had become murderers or run amok, they would have killed a few or a dozen people before apprehension by the police.

Instead, it should be emphasized that the grand expressions of aggressiveness are always within a symbolic framework or, to speak in Freudian terms, that it is necessary to analyze aggression in the framework of the secondary as well as the primary process. What makes the Holy Inquisition, the bloodshed of the French Revolution, Communism, or Nazism so pernicious is that destructiveness is within a symbolic structure. Hitler built a symbolic universe in his dream of a Thousand Years' Empire. For this end he condemned the Jews to concentration camps and gas chambers and his Germans to slaughter on Russian battlefields. In private life a grand inquisitor or commissar may be a charming fellow, fond of children and dogs, and he often is. Within the framework of ideologies, the atrocities involving millions of people develop. Compared with this, going berserk or ordinary murder is relatively harmless. Similarly, organized intraspecific aggression, that is, warfare, is specifically human and based upon the symbolism of political and religious systems, states, nations, and so forth.

To express it in a somewhat different way: the specifically human danger is *the coupling of aggressive instincts with constructs at the symbolic level.* In human aggression, there are indeed primitive biological drives, but we must not forget the other, the symbolic component, which makes human aggression infinitely more dangerous and cruel than the biological struggle for existence in the subhuman world. In still other words: The superego is not only a censoring apparatus, as was emphasized by Freud, but is a tremendously creative force, both in the way of construction and destruction.

Since the supreme manifestations of destructiveness or aggression are linked with the symbolic universes of man, I believe that the conventional interpretation of aggression as being a regression to a primitive stage (like that of a primal horde or a primitive father-child relationship) needs careful re-examination. It appears that the interlacing of primitive emotional and drive patterns with highly symbolical structures presents the problem. So the symbolic framework (the so-called quasi needs) is at least just as important as inborn tendencies or instincts.

Here two other considerations should be discussed. The first may be expressed in one sentence. Suicide is an exclusively human phenomenon; animals do not commit suicide. It appears that self-destruction in man is again intimately related to the symbolic world in which he lives, and which in a certain way has a life of its own. So an individual will commit suicide either if his symbolic world demands it or if his symbolic framework breaks down. The first case refers to the martyr. The behavior of Joan of Arc, Giordana Bruno, or the Christian martyrs would be utterly incomprehensible considered from a utilitarian or homeostatic viewpoint. All of them could have easily saved their biological existence and survival by the simple expedient of renunciation. That they did not do so was due to the fact that their respective symbolic universes were more important to them than their individual lives. In a martyr we have an example of self-destruction. But we do not attribute it to a death instinct or a tendency to self-annihilation; rather we give it a positive value, and consider it the result of the conflict of the symbolic world of an individual with the traditional symbolic worlds in that particular instant of history. On the other hand, the man who kills himself because his life or career or business has gone wrong, does so, not because of the fact that his biological existence and survival are threatened, but rather because his quasi needs, that is, his needs on a symbolic level, are frustrated. A milder form of this is anxiety. Animals may have fear of an immediate danger. But only humans have anxiety of the future, a fear which is based upon the anticipation of future in thought, that is, in symbolic images.

A second point I would like to make is that this symbolic world "cuts both ways." On the one hand, it leads to the most catastrophic manifestations of aggression such as ideological warfare; on the other hand, it may provide a means to check the aggressive tendencies in man. It may do so either by checks imposed on aggression by symbolic structures of conventions, morals, and religion; or by transference from the biological to the symbolic level, as when killing the enemy is replaced by a rather mild form of aggression such as economic competition or playing chess and bridge.

Rousseau's idyllic image of the good natural man, spoiled only by civilization, and Freud's image of man as a born aggressor, patricide, and fornicator, precariously kept in bounds by the Censor, are equally romantic and unrealistic. Man is a tolerably monogamous species, he is a social animal, and he has inborn parental and filial instincts. His instinctual equipment, however, is not specifically human even though it may be the basis for conventional moral codes. Rather it is shared by any number of animal species which exhibit such instincts in far greater perfection.

If we look at the increase of the brain in the series of vertebrates from fish and amphibians to reptiles, mammals, apes, and finally man, we notice an important phenomenon. This evolution is almost exclusively an increase of the neencephalon, the hemispheres (see Appendix). This implies that the so-called progress of man is essentially an intellectual affair, made possible by the tremendous development of his forebrain. Owing to this, over the course of cultural evolution, man was able to build up the symbolic worlds of speech and thought and some progress in science and technology. No development, however, is seen on the instinctual side, for the perfectly good reason that there is no anatomical substratum for it. The human cortex provides some 10 billions of neurons for intellectual achievements, but hardly a basis for a parallel development on the emotional and instinctual side. We have progressed, in some five thousand years, from primitive mythology to quantum theory, and in some 150 years from the steam engine to the hydrogen bomb. But it would be a slightly optimistic view to say that general moral standards have progressed since Lao-tse, the Buddha, or Christ.

However, the so-called neutralization of aggression appears to be on the level of symbolic constructs, which is what all systems of morals, religion, etc., amount to. Although there appears to be little evolution in man with respect to instinctual (or subcortical) functions, still these lower neural apparatus and corresponding instincts may be inhibited by functions at the higher symbolic level. The inhibitory action of higher centers on lower ones is a well-known fact in neurophysiology. It appears, then, that we cannot change the *bête humaine*: we can only hope that the brute in man is better controlled.

This appears to coincide with Waelder's (1956) comments on maturation being instrumental in counteracting primary aggression. Not much can be added to what Menninger (1942) has already said in *Love Against Hate* about the possible means for sublimation and detoxification of human aggression. But, from the viewpoint taken here, the acceptance of adequate symbolic restrictions within the given cultural system might constitute such a means. This, then, would be the "maturing ego" of which Waelder speaks.

4
Systems Perspectives on the Problem of Mental Illness

The basic question to begin with is whether there is a specific "disease" named *schizophrenia* as distinguished from others. I submit that for defining a disease per se, a triad of criteria is necessary: (1) a specific origin or etiology, (2) a specific syndrome of symptoms, and (3) a specific course (or specific ways) of treatment. While such conditions as appendicitis or pneumonia comply well with these criteria, this becomes somewhat more questionable in diseases such as rheumatism or arthritis, where the name is little more than a label attached to disturbances whose etiology is obscure, symptoms vary, and specific therapy is undeveloped.

In the light of the criteria mentioned, apparently schizophrenia does not qualify as a specific disease. Its etiology is unknown; its symptomatology offers a spectrum of aberrations from the normal, which is not only most multifarious but largely divergent in individual cases; and no typical course or specific therapy is known. Thus, I hardly need to say that, from my biological viewpoint, I embrace the "unitary concept of mental illness" as advanced by Karl Menninger and his group (Menninger, Ellenberger, Pruyser, and Mayman, 1958). The unitary concept in psychiatry is closely related to the so-called organismic viewpoint in biology (see Bertalanffy, 1952) which, as opposed to the mechanistic view and approach, emphasizes the organism as a whole and its dynamic organization.

A simile taken from the field of organic disease may be helpful in evaluating the unitary concept of mental illness. Speaking of "cancer," it is obvious that there are many kinds of cancer of the various organs. Histologically there is a great difference between, say, a squamous cell carcinoma, an adenocarcinoma, or a hepatoma. Different therapies such as surgery, as opposed to radiation, may be indicated. The borderline between benign tumors and malignant growth is not clear-cut, and in some cases a benign tumor appears to develop into a cancer. However, even though "cancer" is a multifarious disease, its different kinds are not considered separate "diseases" although, of course, different terms are used in diagnosis. Rather, it is generally agreed that there is some cytochemical and physiological aberration—by no means clearly understood at present—that underlies the various forms of malignant growth, so much indeed that it is unnecessary to speak of a "unitary concept of cancer."

In a similar way, a unitary concept of mental illness will not deny patent differences or contend that mental disorders differ only in degree or quantity. The idea is rather that there is some basic disturbance which may be manifest in different ways. The contrary—the reification of psychiatric terminology into separate disease entities—is as if we were to consider lung cancer, skin cancer, uterine cancer, or liver cancer as separate diseases, instead of different forms of malignant growth.

It deserves mentioning that Eugen Bleuler, creator of the term "schizophrenia," did not commit the fallacy of reification of psychiatric terms. This is apparent even from the title of his classic (Blueeler, 1950), referring to the group of schizophrenias" (i.e., a number of frequently encountered syndromes) rather than to one entity bearing his name.

On the other hand, recent biochemical hypotheses imply that there is some sort of metabolic disturbance which is the cause of schizophrenia, i.e., an aberration of epinephrin metabolism, deficiency or excess of cerebral serotonin, taraxein as a plasma protein causing schizophrenia, increased ceruloplasmin level. Although none of these claims is well-established at present (see Bertalanffy, 1957; McDonald, 1958), probably a "biochemical lesion" (Minz, 1958) is a predisposing factor of psychosis.

However, a one-sided biochemical approach presupposing a uniform disease called "schizophrenia" and a simple cause-effect relationship between an alleged toxic agent and mental disease (similar to some infection causing pneumonia) is apt to lead to disregarding the variety of etiological factors and the diverse course and prognosis of mental illness.

The unitary concept of mental illness, propounding that mental disorder is a *systemic* disease rather than a number of separate diseases, I believe represents a major breakthrough in psychiatry, which may lead to a basic reorientation in theory and clinical practice. The search for separate entities of mental disease should be replaced by the concept that the delicate mental apparatus can be disturbed in various of its parts and to a varying degree. As in every disease, the goal of therapy should be the re-establishment of the normal organismic order, which has been called the *vis medicatrix naturae*. Then, not only can otherwise paradoxical facts be intelligently arranged, but we may also concentrate on etiological research and therapy which is comprehensive and not one-sided, and thus more promising than previous efforts.

Biochemical Theories and Model Psychoses

If we accept the unitary concept of mental illness, a number of consequences follow, which throw light upon and entail precautions with respect to a number of recent approaches. The first is the question of whether there is a biochemical basis for schizophrenia. The British psychoanalyst Sandison (1957, p. 19) has characterized the present state of affairs as follows: "North American opinion seems to favor the old medieval idea of a drug to counteract an alleged toxin (i.e., the search for a hallucinogenic substance circulating in schizophrenics and research into antihallucinogens)."

In all probability there are biochemical and physiological disturbances in the brain mechanism, possibly linked with genes, which are connected with mental dysfunction. However, it seems questionable whether, at the present level of investigation, biochemical disturbances specific to schizophrenia can be found. It would probably not be difficult to find in the literature a dozen or more biochemical tests for schizophrenia which, at the time, were enthusiastically welcomed. Their general nature, however, is similar to that of a more recent one: the ceruloplasmin test. The concentration of this copper-containing enzyme is supposedly increased in psychiatric patients. However, a similar increase was also reported in cases of liver disease, infection, pregnancy, etc. (Akerfeldt, 1957; Gussion, Merle, and Kuna, 1958). That is, biochemical differences may be detectable, but their specificity is problematic.

Similar considerations apply to the number of biochemical theories proposed in recent years to provide a physiological basis for schizophrenia (Bertalanffy, 1957). Without entering into a detailed discussion, it seems that a critical attitude is indicated at present. Reinvestigations at the National Institutes of Mental Health of some major biochemical hypotheses on schizophrenias (erythrocyte glutathione level, Akerfeldt test, adrenochrome, serotonin, taraxein) failed to confirm the claims made (McDonald, 1958). A considerable number of other investigations to a similar effect could easily be cited. Apart from factual criticism, the question of the explanatory value of biochemical theories should not be overlooked. Suppose, for example, that disturbances of the catecholamine metabolism or of the normal serotonin level in the brain were safely established as concomitant with psychosis (as, apparently, they are not at present). We may still ask what this means with respect to the clinical symptoms named schizophrenia: In what way is this connected with disturbance of association, split personality, delusions, hallucinations, and the rest of the schizophrenic symptoms?

Recent biochemical theories of psychosis and of schizophrenia in particular are largely based upon the so-called model psychoses. The basic question then, is whether states induced by the hallucinogens (with mescaline and LSD as foremost representatives) are a closer "model" of schizophrenia than those caused by more trivial agents, which range from alcohol, opium, hashish, carbon dioxide (CO_2), metallic poisons, to sensory deprivation (Bindra, 1957; Lilly, 1956). On the basis of such comparison, the psychotic states induced by mescaline and lysergic acid (LSD) appear to be symptomatically closer to toxic psychoses than to clinical schizophrenia (although with relatively harmless vegetative disturbances).

New Concepts in Human Behavior

A biological, as distinct from a biochemical approach, will have to take into consideration the various aspects of the human being and of mental illness. A question the biologist is bound to ask and in some way to answer offers a suitable starting point. Namely, what is characteristic of human as compared to animal nature and behavior? As the psychiatrist is concerned with disturbances of normal psychology and behavior, this question obviously is a basic one. There are two concepts which I wish to offer in this respect: namely, first, man as an intrinsically active, psychophysical organism (see Chapter 9); and second, man as an *animal symbolicum* (see Chapter 1).

The first concept deserves brief explanation. According to the classical model of behavior, namely the stimulus-response scheme, an organism (man included) functions by way of answering stimuli coming from outside. Behavior is essentially directed toward the re-establishment of an equilibrium disturbed by external factors, and hence it is essentially governed by economic principles. Thus, behavior is essentially a sum total of reactions tending to re-establish, at minimum cost, a supposed "equilibrium" which was disturbed by outside stimuli. Similar considerations apply to psychological and psychoanalytic theory. The stimulus-response scheme underlies Freud's "principle of stability," according to which the supreme tendency of the organism, biological and mental, is to get rid of stimuli and come to rest in a state of "equilibrium." The same applies to the more recent model of feedback and cybernetics which still adheres to the stimulus-response scheme—only the circular feedback loop is added. The principle of economy is expressed by the notion that behavior, infantile and adult, essentially means "coping" with reality, i.e., dealing with environmental demands and challenges at minimum expense.

Although these principles have proved useful in psychological and psychoanalytic theory and practice, it appears time to expand them. I believe that a more basic reorientation is necessary, one which, as a matter of fact, can be supported by a large amount of biological, neurophysiological, behavioral, psychological, and psychiatric evidence. This essentially new notion (new, however, only with respect to the conventional stimulus-response scheme) is that of the primacy of immanent activity of the psychophysical organism, a notion related to the theory of the organism as an *open system*. Even without external stimuli, the living organism is not at rest, but is an intrinsically active system. This is evidenced, for example, by the fact that *active* behavior phylogenetically and ontogenetically precedes *reactive* (reflex) mechanisms which are superimposed as a secondary regulatory apparatus.

Furthermore, the living organism does more than "maintain its equilibrium." As long as it lives, it maintains a disequilibrium called the "steady state" of an open system; and it even advances, ontogenetically and phylogenetically, to higher forms of order and organization (that is, to states more distant from equilibrium).

Similarly, behavior not only involves reaction to outside stimuli and exigencies at minimum costs, gratification of needs and "equilibration" with the environmental situation; the activity of the organism also includes what may be loosely called a creative element, where "function pleasure"—as Karl Bühler (1929) called it—which accompanies the activity is its own reward. Moreover, not only conditions of stress, but equally those of absence of stimulation are psychopathogenic factors. In the laboratory, sensory deprivation in an isolation chamber soon leads to a model psychosis. This apparently is related to well-known clinical phenomena such as prisoners' psychoses and the exacerbation of schizophrenic symptoms by isolation in the ward. At the level of social factors, the insight is increasing (in existential analysis and beyond) that meaninglessness and emptiness of life, rather than Freudian frustration of biological needs, have become foremost psychopathological factors. Conversely, gratification by self-rewarding activities and not mere gratification of basic biological needs underlies the beneficial effects of occupational and adjunctive therapy.

All these can be considered consequences of the fact discussed above: that the psychophysical organism not only reacts to stimuli coming from outside but rather is a system tending to maintain its intrinsic activity. At the level of human psychology, this has its counterpart in a drive toward "self-realization" with respect to the gratification of both biological needs and those arising within a symbolic system of values characteristic of a certain social and cultural framework. If this value system becomes problematic and if the individual does not find orientation and satisfaction in his way of life, it is not to be wondered that meaninglessness of life may be as potent a neurosogenic factor today as was repression of instinctual drive in Victorian times.

It appears that it is not conditions of biological stress, but rather of stress at the symbolic level—"quasi-needs" (Hacker, 1958; Lewin, 1935) that lead to an increase in the rate of mental disorder. It has repeatedly been noted that World War II, a period almost unparalleled in physiological hardship (a daily ration of often 1,000 calories or less per day) as well as psychological stress, did not bring about an increase either in neuroses (Opler, 1956) or in psychoses (Llavero, 1957). Conversely, in the United States today with an economic opulence never reached in the past and, as far as material comfort is concerned, enjoying a period when "the greatest happiness of the greatest number" has been realized in an unprecedented way, there has been an exasperating increase in mental illness. In other words (apart from easily understood phenomena like combat neuroses), it appears that conditions of biological stress, where animal survival is at stake, are not necessarily psychopathogenic; whereas conditions involving symbolic values or "quasi-needs" (such as money, status, position, comfort, and the like) may be.

This is hardly compatible with theories which seek to explain mental illness exclusively in terms of primitive animal drives and their frustration. It is, however, in perfect agreement with the concept of man as *animal symbolicum*, essentially living in an ambient of symbolic values, seeking fulfillment of these, and being victimized by their frustration.

Application to Psychopathology

These concepts serve to remind us of certain characteristics of schizophrenia which are apt to be overlooked in the too zealous search for a biochemical or physiological theory of that disease. For instance, it seems compatible with the unitary concept of mental illness (but hardly with the concept of schizophrenia as a distinct disease) that, in the very exaltations of the human mind, experiences occur which psychiatrically would be termed schizophrenic symptoms.

Several observations present themselves in this connection. First, schizophrenia, except for the progressive deterioration in dementia, is not simply a degeneration of the mental apparatus but contains creative elements. If the schizophrenic builds up a world-picture of his own, this is a creative act—be it ever so misguided and fantastic from the viewpoint of the "normal" observer. This creative element is the link connecting the schizophrenic and the artist, mystic, and even the scientist. The ideas, for instance, that there are antipodes, that the earth revolves around the sun, that non-Euclidean geometrics apply to physics, or that physical matter for the most part consists of holes or empty space, contradict all common sense and would be considered schizophrenic (as historically they were) if they did not happen to be scientifically correct.

Furthermore, schizophrenia is an essentially human disease because its basic symptoms, "loosening of associational structure" and "splitting of personality" (according to Eugen Bleuler [1950]), are intimately connected with the symbolic activity characteristic of man. Animals may behaviorally show and experience perceptual, motor, and mood disturbances; however, they cannot display a schizophrenic disturbance of *ideas* such as delusions of persecution and grandeur since there are no ideas for them to begin with.

Lastly, one should consider what is meant when "split personality," "losing of ego boundaries" and "self-identity" are regarded as the basic disturbances in schizophrenia. The normal experience in our Western culture is characterized by firmly drawn ego boundaries, "ego" vs. "non-ego"; Descartes' "I think, therefore I am" and the opposition of a *res cogitans* and *res ex-*

tensa are its classical conceptualizations. This "normal" experience of the self and an outside reality gradually develops in the child in conjunction with the development of symbolic categories of thought (see Bertalanffy, 1956); however, the normal universe of the adult European is not the only possible one. A "loosening" or even "losing of ego boundaries" appears to be the common feature of forms of experience which are different from the "normal" but are not necessarily "pathological," for example, the animistic world-picture of primitives, the supernormal states of "peak experience" (Maslow, 1959), the mescaline or lysergic acid experience, and psychotic states. One may also wonder whether the Buddhist's "That art thou" does not express a very different world experience wherein empathy (a fundamental nuisance in a physicalistic world view) plays a role hardly accessible to us.

Thus it appears that the structuring of experience considered as normal as well as deviations from it largely depend on one's cultural background. Consequently mental disturbance means not only distortion of sense data (as in hallucinations, illusions) but to a large extent a distortion of culturally accepted symbolic interpretations.

If such symbolic interpretations appear particularly unusual, we call them by the psychiatric name of "delusion." Or else, there may be a "concretistic inability to symbolize" or "loss of symbolic capacity" (Goldstein, 1940; Hacker, 1958), that is, an unrealistic mistaking of symbols for "things" and vice versa. Reification of concepts, that is, making mental images into real things, is one characteristic of schizophrenia; but the same trend is found in much of our normal thinking, the borderline being indistinct. It ranges from the illustrious example of Plato turning abstractions into reality (and, indeed, into the ultimate reality), to the anthropomorphism of early physics making "forces" (a simile taken from introspective experience) into an outside reality, only laboriously replaced by a "deanthropomorphised" world-picture of purely mathematical relations (Bertalanffy, 1955), and includes popular notions like "nation," "humanity," and the like, collective nouns for social groups, which are made into emotion-charged fetishes.

Examples of this kind, which could easily be amplified, lead to the question whether and to what extent mental illness is connected in its epidemiology and its content with a given cultural framework (Opler, 1956). For instance, Lambo (1957) reported that in the native population of Africa manic-depressive patients preponderate in comparison to schizophrenics. Epidemiologic shifts are well known in our own society which are of a short-range nature so as to exclude the possibility of genetic differences. One example is the near disappearance of classical Freudian hysteria and its replacement by obsessional neuroses or psychosomatic disorders. One may speculate whether the recent increase in schizophrenia is not promoted by a certain cultural situation, namely, the conflict of the traditional Western ego ideal with the impact of "other-directedness" (Riesman, 1950) and the "organization man" (W.J. Whyte, 1956) in modern society.

Taking into account this "culture-boundedness" of behavior, there appears to be only one criterion which distinguishes mental health from mental illness. It is that the mentally healthy individual has a consistent and integrated universe, while the schizophrenic has not. Hence similar symptoms may appear in the supernormal experience of the genius and the mystic, in primitives, and in the schizophrenics of our hospitals. So whether an individual is considered "sane" or "insane" is determined not by isolated symptoms only, but rather by whether these are embedded in an organized universe consistent with the given cultural framework.

5
A Definition of the Symbol

The Problem

What Distinguishes Man

Modern students of behavior generally agree that symbolic activities distinguish human from subhuman behavior. Indeed, symbolism so pervades human activities that one wonders why study of this obvious aspect of behavior and psychology has been for so long (and is even now) neglected.

However, for a long time *language* was considered the distinguishing characteristic of man. This characterization is unsatisfactory for three reasons. First, there is the question of animal languages which obscures the distinction if no qualifications are made. Second, the concept of symbolism is far broader than that of language, including, as it does, a main feature of "cultural activities" in general, which certainly are germane to man. Third, language is a highly developed form of symbolism rooted in deeper and more primitive layers where the junction between subhuman and human behavior may possibly be found.

Another definition is that man is a *rational animal.* That definition is hardly palatable to a generation which went through two senseless world wars and doesn't need the testimony of dynamic psychology to realize what a small part of human behavior is "rational." With due deference to game theorists and mathematicians (to whom "rational" decisions are an axiomatic basis for their elegant deductions), we may justly say that man is a symbol-creating, symbol-using, and symbol-dominated animal throughout. But the use he makes of this remarkable achievement is only in a minute part rational. Symbols, such as mink coats, flags, anthems, television advertisements, political catchwords, deterrence, and what not, determine individual and social behavior only too often to the detriment of the individual, of society, and of humanity. If we take "rationality" to mean behavior favoring maintenance of the individual and species and full realization of their potentialities, then animal behavior is indeed much more "rational" than that of *Homo sapiens*.

Still another classical definition is of man as *Homo faber*. Man the Engineer is a criterion that must appeal to an era in which engineering activities have become dominant. Nevertheless (Köhler's experimenting apes notwithstanding) there is a suspicion that symbolic activities, thought and language, precede the making of tools. Without anticipation of the goal in thought, possible only on the basis of symbolic representation, man would hardly have advanced beyond a neolithic stage when stones, suitably formed by natural accident, were used as hammers and axes. Only some foresight of what to achieve could have led to a purposeful preparation of tools. *Homo sapiens* probably antedated *Homo faber*, if such distinction of closely allied faculties is at all permissible.

The Need for a Definition

In spite of the fact that symbolic activity is one of the most fundamental manifestations of the human mind, scientific psychology has in no way given the problem the attention it deserves. Furthermore, there is no generally accepted definition of "symbolism."

We are, therefore, in a somewhat paradoxical situation. There is general agreement that "symbolism," however defined, is a basic attribute of human behavior, being precisely what distinguishes man from other beings. This consensus extends from evolutionary biologists such as Julian Huxley, Dobzhansky (1959), and the present author, to students of animal behavior, including those specialized in anthropoid behavior (see Langer, 1948, pp. 84ff.), Pavlovian neurophysiologists studying the "secondary signal system" (Luria, 1961), and anthropologists, philosophers, and others. On the other hand, the term "symbol" does not even appear in the index of leading psychological texts. This amounts to saying that one of the most distinctive criteria of human behavior remains unrecognized and is bypassed by contemporary psychology.

Furthermore, there is wide divergence as to what "symbolism" really means. The spectrum of meanings ranges from Carnap's "logical syntax of language," mathematical logic, and general semantics, to Goethe's profound concept of symbol and Spengler's *ur*-symbols of culture, to Freudian and Jungian symbols and the religious symbols of which Tillich speaks.

There have been comparatively few attempts to give the problem the place it deserves. The first was Cassirer's monumental work (1953-1957); another was Susanne Langer's book, *Philosophy in a New Key* (1942); these should be the "bible" to the student of symbolism, to be taken as the basis for any further discussion. The present author may freely acknowledge this because he developed his own related formulations (Bertalanffy, 1947) independent of either work, at a time when in Austria (then his residence) there was no possibility even to know about Langer's book.

The above considerations illustrate the difficulty and multifariousness of the problem. We speak of the symbolism of mathematics, physics, and genetics, as well as of music and painting. We have status symbols. Many will agree that specific human values may be distinguished from general biological values (such as adaptation of the individual and survival of the species) by the fact that they are at the symbolic level. Economic values are only too predominant in an Affluent Society, but there is also much talk about values of religion and, almost in the same breath, of Freudian sex symbols. In this array of different notions, what is the common denominator that would permit a definition?

The authors who, as pioneers, have laid the foundations for a theory of symbolism, have done so by way of description and illustration rather than definition. This is understandable and necessary. If we are to explore entities, such as plants, animals, or concepts, in a newly discovered continent, the first task is collecting and describing. The next step will be definition and taxonomy. Even if only tentative, this process will in some way circumscribe the species to be studied, and so yield a basis for further exploration and research.

A Definition of Symbolism

The confirmation of any theory is in its application to concrete phenomena. The theory of symbolism to be outlined has been applied to a number of otherwise unconnected facts and was found to be of explanatory value; otherwise disparate phenomena fall into place "like pieces in a jigsaw puzzle." For example, the author has applied this theory to such phenomena as the distinction of human and subhuman behavior (see Chapter 1); problems of human values, of modern society, and of education (see Chapters 2 and 11); psychopathology (1966a, 1968, 1969, 1971, 1972; see also Chapters 3 and 4); child psychology and mental growth (1956); the action of hallucinogenic drugs (1957); the categories of human knowledge and the evolution of science (1955).

The definition of symbolism which was proposed earlier (see Chapter 1) was especially intended to distinguish human from subhuman behavior. This distinction was said to be satisfactorily achieved when symbols are defined by three criteria jointly applied; that is, when symbols are defined as signs which are (a) freely created, (b) representative, and (c) transmitted by tradition. Taken together, these criteria appear to be necessary as well as sufficient to distinguish human symbolism, and in particular, human language, from subhuman forms of behavior.

If symbols are defined as "freely created," we mean that there are no biologically enforced connections between the sign and the thing connoted. This provides a distinction from conditioned responses. In the case of conditioned reactions the connection between signal and thing signalled is imposed from outside. This may be a natural connection as when the sight of a flame leads to a motion to avoid the fire because the child or the kitten has burned itself in a previous instance. Or else the connection between signal and thing may be imposed by the experimenter,

Definition of the Symbol

as is the case in Pavlovian experiments: The dog secretes saliva after the ringing of the bell because ringing of the bell was regularly followed by a meal during the conditioning period. In this sense, the connection between signal and thing signalled in conditioned reaction, in both the Pavlovian and Skinnerian senses, is "imposed from outside." On the other hand, there is apparently no biological connection between the word "father," "Vater," "père" (or whatever the word may be in any language), and the person so designated; therefore, symbols are said to be "freely created."

The second criterion of the definition of symbolism is closely connected with the concept of language. The renowned Viennese psychologist Karl Bühler (1929, 1934) distinguished three functions and corresponding types of language: language *as expression*, as *elicitation*, and as *representation (Kundgabe, Auslösung, Darstellung)*. Now, "language as expression" is not a specifically human phenomenon. For example, a bird's song certainly expresses and communicates to his mate a certain physiological and, we may be sure, psychological state; but it does not connote a thing. Language as elicitations or command is also not specifically human. Certain gregarious animals by acoustic utterances warn the herd of an impending danger; this is of the same nature as the officer's command, *attention*, which does not communicate a content but elicits a certain action. Again, the representative aspect is lacking; the barking of a dog warns of some danger, but it does not communicate what the danger is, whether an intruding burglar or just the neighbor's cat. In contrast, human language—and symbolism in general—is "representative."[1] Representativeness is also characteristic of non-linguistic symbols, for example, a stone representing commodities in economics, flags or slogans representing a political entity.

Finally, symbolism and human language are defined as being transmitted by learning and tradition. This criterion is important to distinguish human speech from animal languages. For example, the language of the bees studied by von Frisch certainly is "representative"; by means of intricate dances, bees do communicate to their colleagues where and what food can be found. However, this language, innate and instinctive, does not fulfill the criterion of transmission by learning. We can teach a dog all sorts of tricks but we have never heard that a particularly clever dog has taught her puppies to do them. If, on the other hand, tradition *is* found in animal behavior—such as in certain bird songs or in the famous example of the appearance of a "subculture" in titmice which have learned to open milk bottles—the other characteristic of a "meaning" or being "representative" is found lacking.

The point to be emphasized is that all three criteria are necessary to delimit symbolism from subhuman forms of behavior. We do find animal behavior and languages fulfilling one or the other of these criteria; however, we do not find all of them together except at the human level. Even in anthropoids this form of activity is not found. As a matter of fact, the most competent observers commented on what they called the "aphasia" or "mutism" of chimpanzees (Langer, 1948, pp. 94ff). This condition is probably connected with the evolution of the brain. The sensory speech center localized in the temporal lobe is quite well developed in certain animals; hence, the dog, the elephant, and the horse can learn to respond to verbal commands and so, obviously, have acoustic gnosia of speech. In contrast, the motoric speech center in the frontal lobe seems to be lacking in animals, even in anthropoids, and seems to appear only in man (Rensch, 1959).

Whereas the latter two of these criteria, representation and tradition, are commonly accepted, I have often encountered objection to the first, "free creation."[2] "Freely created" in the sense of the definition does not imply "voluntarily, arbitrarily, consciously, or rationally produced," although some of these characteristics do apply to some symbolic activities. It is to be taken strictly in the antithetical sense stipulated, that is, in contrast to "biologically imposed" or "necessarily of biological value." Symbolism may or may not have biological and adaptive value. It has adaptive value in the case of science and technology (with qualifications obvious in our time of troubles). However, a so-called homeostatic value can hardly be ascribed to Greek sculpture or German music, either for the individual or for the society concerned. And such symbolically determined behavior as suicide or war is definitely disadvantageous from a biological standpoint. It is in no way denied—and we shall come back to this—that there are connections between the biological and symbolic, subhuman and human levels. In the origin of language certainly biological sounds (expressive outcry) and onomatopoeia (mimesis in words) play an important role. Apart from direct onomatopoeia, words similar to "mama" and "papa" are found with identical meaning in the most diverse languages; apparently they originate in the babbling of the baby (see Jakobson, 1960). Similarly, imitative behavior is at the root of much symbolic representation in ritual. Nevertheless, the criterion remains valid even if biological material is used for "freely created" symbols.

The definition and classification of symbolic activity of man, of course, does not imply that it is an isolated function. Various classes of symbols are not neatly separable; intermediates and interactions abound. The same is true with respect to other cognitive (e.g., perception, learning) and non-cognitive functions. Whereas the criterion of "free creation" applies to the origin of symbols, learning and using existing symbol systems ("verbal behavior," Skinner, 1957) is determined by operating behavior—excepting (we may roughly say) creative recombination of existing symbols.

Moreover, there are obvious connections of symbols with the affective aspects of behavior. Animal ethology speaks of "instinct-training interlocking" (*Instinkt-Dressur-Verschränkung*, Lorenz, 1935), indicating that a close interaction between innate and learned components is often found in animal behavior. Certain components in, say, nest building or flying, may be innate neurophysiological mechanisms; others have to be acquired by learning. Introducing a somewhat similar term, we may speak of "symbol-affect interlocking" in order to indicate an important and often pernicious interaction between symbolic and emotional components. Symbols, particularly those of a non-discursive nature, often enter into intimate connection with higher emotions or with animal drives. This linkage may lead to the highest elations of the human mind in artistic, moral, religious, even in scientific experience. It may also lead to coupling the symbol with the basest tendencies of the human animal. The sublime symbols of religion may be coupled with sexual frustration or aggressive drives, leading to witch-hunting and liquidation of heretics *ad majorem Dei gloriam* (the god in the formula being replaceable by any pseudo-religious symbol or fetish). As a matter of fact, such emotional influences are primarily responsible for making the "spectacles" of symbolism into distorting lenses which basically change and falsify man's outlook of the world—which replace what psychoanalysts call the reality principle by delusions and illusions that often approach the schizophrenic level.

The animal is safe because it is living in the innate niche or shell of its instinctive *Umwelt*; the sage is safe within his universe of freely created symbols. Between them is the battleground of biology and psychology where man, owing to his specific endowment, behaves more beastly than the beasts. Symbols become idols, overriding and often destroying the individual or society and openly clashing with the biological values of self-preservation and survival, the values of "rationality," "homeostasis," "adaptation," and "adjustment to reality."

Further Characteristics of Symbols

Two further characteristics of symbolic universes are what may be called metaphorically their "productivity" and "autonomous life" (see Chapter 1). Symbolic systems, systems of discursive symbols (language, mathematics, science), are "productive," i.e., they yield more "dividends" (Rapoport, 1955) than the conceptual capital originally invested. Owing to their grammatical structure they develop into an algorithm or "thinking machine" which permits prediction according to Hertz's dictum that "the consequences of the pictures are the pictures of consequences."

Furthermore, symbolic universes follow autonomous laws, i.e., laws of their own which are not identical with or reducible to laws of biology, psychology, sociology, and so on. The evolution of Renaissance painting from Giotto to Tintoretto, of music from Gluck to Wagner, of the Roman law have, as it were, a logic or life of their own, transcending the psychology of the human individuals who created them. This is what Hegel

Definition of the Symbol

meant by "objective spirit" and its evolution. Sometimes it is possible to formulate definite laws of the evolution of symbolic systems, for example, Grimm's law on the successive changes of consonants in linguistic development.[3] However, more often, it is not possible.

Discursive and Non-discursive Symbols

Classification of the universes of symbols implies an immense task; only a few incidental ideas can be offered.

We would tentatively propose to define discursive symbolism or language in the wide sense as communicated information which in principle (although hardly in fact) is measurable in bits. This covers, first, languages with a vocabulary but no grammar, such as the flag language (e.g., Paul Revere's message) or the language of the African tom-tom; second, propositional languages which not only have a vocabulary but also a grammar and therefore are capable of forming the basis of algorithms that permit prediction. The second applies, to a degree, to vernacular language; it applies in the highest degree to the artificial language of mathematics.

It may be mentioned in passing that mathematics is not confined to a "science of quantities" or metrics and that mathematics in the conventional sense is not the only deductive and predictive language. Modern branches of mathematics, such as mathematical logic, topology, game theory, etc., are not metric although they are deductive. Less esoteric are the languages of chemical and genetic formulas and of taxonomy, which are outside conventional mathematics but are deductive-predictive in varying extent.

The second great realm of symbolism is much harder to approach and is essentially a new field for analysis. Cassirer and Susanne Langer here have laid the groundwork. We may tentatively characterize this field as that of "non-discursive" symbolism:

> There is an unexplored possibility of genuine semantics beyond the limits of discursive language. This logical "beyond" which Wittgenstein calls the "unspeakable," both Russell and Carnap regard as a sphere of subjective experience, emotion, feeling,

and wish. The study of such products they relegate to psychology, not semantics, and here is the point of my radical divergence from them. The field of semantics is wider than that of language. (Langer, 1948, p. 70)

To give a brief example: A lyrical poem containing the words "sunshine" and "spring" does not, as Carnap would have it, inform us about meteorological data; neither is it mere expression of feelings as an (unnecessarily) stylized outcry "Ah-Ah." Rather, it communicates the intimate experience of Goethe. These, in contrast to discursive symbols, may be referred to as *experiential* or *existential* symbols.

Although these experiential symbols are beyond the field contemplated by epistemologists of the positivist school, they nevertheless follow the criteria for symbols enumerated earlier, being freely created, representative, and bound to tradition. Music, for example, is a symbolic "language" not for communicating thought, but for communicating intimate experience. It has to be learned, as shown by the trivial fact that, say, Chinese music is not intelligible to us and that "modern" music of any period is first resisted as being unintelligible. The same is true of discursive thought, when, for example, the existence of antipodes or the Copernican system was first resisted because it was opposed to "common sense," and when the insights of science, fantastic as they are in comparison to direct experience, eventually are taken for granted and are regarded as inoffensive.

With non-discursive symbols, we are moving (as Langer has emphasized) at the very limits of the ineffable; it is precisely for this reason that the realm is so hard to define. The realm of "non-discursive" symbolism is at present a catch-all which requires much elaboration. In the field of discursive symbolism we may distinguish between everyday language and higher technical languages such as mathematics (typical products of "high culture"). So, too, in the non-discursive field, we can find similar levels.

On the everyday level, the non-discursive category should include everything that can be communicated symbolically but cannot be expressed in discursive, yes-or-no type answers or bits (applying the terminology of information theory). At this level belong symbols referring to the individual (status symbols, for example); symbols referring to a social group such as a company, army, or nation (insignia in the broad sense); economic symbols such as money; and many others. Symbols such as banners, anthems, democracy, communism and whatnot cannot be verified by discursive or scientific thinking. They are expression and communication of personal or social experience, culturally determined in space and time.

Although it would require much more exploration than is possible here, it would appear that whereas discursive symbols convey *facts*, non-discursive symbols convey *values* felt and acted upon, that is, emotions and motivations. These values are, in general, *human values*, that is, values that transcend the biological values of maintenance of the individual and survival of the species. This relationship becomes clear when there is a conflict between biological and human values. Being an executive vice-president with ulcers is not necessarily the optimal biological, psychological, or social "homeostasis"; national or army prestige leading to nuclear war may be not only deleterious for many individuals, but also of dubious survival value for the state, nation, and human species.

The higher level of non-discursive symbolism is that of culture in art, poetry, music, morals, religion, and so on. We may, perhaps, call it "existential symbolism." One will at least come near to what is meant by adopting Maslow's term (1959) and saying that these are forms of representation and communication of "peak experiences." As is well known, Maslow has distinguished between the normal "deficiency cognition," that is, cognition in the service of adaptation and coping with reality, using adaptive perception and accepted symbolic frameworks of discursive thought and language, as different from "being cognition" attained in the peaks of love and of mystic, esthetic, orgasmic, and other experience. "Peak experience" is non-utilitarian, transcends the boundaries between ego and non-ego, renounces rubricizing (in the terminology here employed: bringing things into the framework of discursive categories), and is detached from personal goals and anxieties. Conventional Western psychology, however, takes only "deficiency cognition" into account.

Maslow's terms are linguistically awkward and do not fully convey the meaning. Nevertheless, definition of non-discursive symbolism as representation and communication of "peak experience" is probably the best we can do. This form of experience is "ineffable" in the sense that it cannot be conveyed in discursive categories and language which—as Maslow correctly implies—are essentially a means to successful intercourse with the ambient universe. It can be conveyed in the way of metaphor or simile, as in art and poetry; in non-discursive symbolism such as music; and in symbolic systems of values, as in morals and religion. If the meaning of Goethe's *Faust*, of Van Gogh's landscapes, or Bach's *Art of the Fugue* could be transmitted in discursive terms, their authors should and would not have bothered to write poems, paint, or compose, but would rather have written scientific treatises.

In the ultimate "existential" experience even these symbols prove inadequate. Because language is essentially constructed for coping with the ambient, for "deficiency cognition," and for utilitarian purposes, the mystic knows his experience to be ineffable, communicable only by metaphors which always remain inadequate. Eventually he must fall back to what in medieval terminology was called negative theology, that is, one can speak of ultimate reality only by way of negations. Here, then, is the limit of symbolism, not only of the discursive and utilitarian kind, but of non-discursive as well.

Associated Terminology

A brief table of terms as used in the present study may be of value here.

Classes of Signs

Under the term "sign" we understand stimuli in the broadest sense, which are effective not by simple physiochemical and physiological action (such as light, sound waves, temperature, gravity, etc.) but by their "meaning," being representative of some other entity for which they "stand in."

1. *Signals:* Are natural or artificial releasers of a *conditioned*

Definition of the Symbol

response; they are either straightforward physiological stimuli or *Gestalten*; they receive their meaning by a learning process during a conditioning period. Examples: the ringing of a bell as a signal for subsequent meal or punishment in Pavlovian experiments; operant behavior (after Skinner); light as a signal for a burning fire in natural conditioning.

2. *Schemata:* Are triggers of instinctive actions; generally of *Gestalt* character, although often very simple; they have meaning by virtue of inherited neurophysiological mechanisms (IRM or innate releasing mechanism, Lorenz). Examples: a certain image, often very simple, characterizing a bird as mother, companion, sexual partner, potential enemy, etc.; simplified faces releasing the young baby's smile; dances as in the "language" of bees; the physiognomic "understanding" of other people's emotive status (shown by experiment to be based upon simple schemata).

3. *Symbols:* As mentioned in the previous section, are signs characterized by being representative, freely chosen, and transmitted by tradition; they receive their meaning originally by a free act of creation, although they often utilize material available from other psychophysiological processes ("presymbols," such as expressive cries or body motions, imitative sounds and movements, depth-psychological associations, possibly others), later transmitted by the usual learning process (e.g., teaching and learning of a language, a formal dance, established rules in art). Symbolism has laws of evolution which are supra-biological. Examples: Grimm's law (of consonant mutations in the historical evolution of Indogermanic languages), "inner logic" transcending the psychology of the individuals concerned in the history of painting, music, laws, science, etc. Symbols are (a) *discursive* (language in a broad sense, including technical languages of mathematics, etc.); (b) *non-discursive* (myth, art, customs, rituals and their material signs, etc.).

Language

The term "language" in the broad sense overlaps the classification of signs given above. Languages can be expressive, elicitative, or representative.

1. *Expressive:* manifestations of inner states of the living being concerned; either innate (e.g., bird's song, cries, smiles of humans), belonging to schemata as described above; or freely created and bound to tradition (music, lyrical poem), then using symbols.

2. *Elicitative:* arousing actions in fellow beings by way of warning, command, etc. Again based either upon innate signs (warning cries of animals) or conventional signs or symbols (red for stop, green for go). In the latter case, acquired by the usual learning processes.

3. *Representative:* acoustical, optical, etc., symbols "standing in" for other entities. They may (a) represent things or events by means of symbols that are independent of each other and so form a "vocabulary" (e.g., flag language, if each individual flag communicates only a single event and these can be combined at will); (b) have pre-established rules for the combination of symbols, a *"grammar,"* permitting derivations which appear to be "new" in comparison to the original complexes of symbols. In this case, the language becomes *propositional* and *discursive,* permitting development of a "thinking machine" or algorithm—in the most highly elaborate case, of a "hypothetico-deductive system" (scientific explanation and prediction).

According to the three functions of language (representation, expression and elicitation) a linguistic sign may be either a *symbol* by virtue of its coordination to things and events, a *symptom* as indicating the inner state of the speaker, or a *signal* as controlling exterior or interior behavior of the receiver (K. Bühler, 1960).

Although the concepts as defined are sharply distinguished, actual behavioral events are not. The distinctions can only be applied functionally, that is, with respect to their status within integrated behavior, and depending on the aspect of behavior envisaged. It is therefore moot to ask questions such as at exactly what point (e.g., in the evolution of hieroglyphic and cuneiform writing) a pictogram (of a sun, star, or man) ceased to be an "imitative sign" and became a "freely created symbol." Are the posters for automobile traffic, which by simple drawings indicate "Winding Road" or "Children Crossing," commanding "signals" or representative "symbols"? What precisely is the borderline between "representative" and "abstract" painting; with what degree of artistic interpretation of everyday vision does a painting cease to be "naturalistic"? All such questions are unanswerable for the simple reason that actual behavior is not pigeonholed in neat categories; the distinctions made (or any other classification) are purely operational, that is, intended to facilitate understanding and discourse of phenomena.

Summary

The delimitation and definition of the realm going by the name "symbolism" is required as a prolegomenon to further research.

A definition of symbol according to three criteria—symbols being "freely chosen," "representative," and "transmitted by tradition"—permits us to distinguish specifically human psychology and behavior from subhuman activities. The fields of discursive and non-discursive symbolism were also distinguished and discussed. The latter, because of their connection with processes at the unconscious level, present special difficulties to definition. Tentatively, it is proposed that discursive symbols are concerned with facts, whereas experiential or existential symbols are concerned with values.

Notwithstanding the fact that symbolic activity is generally acknowledged to be a principal characteristic of human behavior, it has, in modern psychology, not found the attention it deserves.

6
An Etymology of Symbolism

A Mini-History of the Concept of Symbolism

The notion of "symbolism" is gaining importance in modern anthropology, philosophy, and other fields. It is rather certain that the origin of the idea is in German philosophy. To quote a few examples, we may refer to the central role that the "symbol" plays in Goethe's philosophical world view and poetry (Bertalanffy, 1949); also, to Oswald Spengler's notion of the *ur-symbols* of cultures (Spengler, 1939, and Bertalanffy, 1924); and to "symbol" as a basic notion in the Jungian version of psychoanalysis. Many other examples could be added.

In contrast, the conception of symbol was neglected or suppressed in Anglo-Saxon philosophy, psychology, and general anthropology. The still predominant neopositivistic school regarded symbolism only in one respect, namely, the discursive symbolism of language, logic and mathematics. This aspect was, of course, extensively explored in logical positivism; the work of Carnap and Morris may be mentioned as representative. The quantitation and theory of information, originated by Shannon and Weaver (1949), may perhaps be regarded as the most important offspring of this orientation although information theory as a mathematical discipline has little direct connection with philosophical neopositivism as represented by Feigl, Hempel, and others. On the other hand, positivistic philosophy of science made a monopolistic claim and hence led to the disregard of all other aspects of symbolism and its basic role in human culture. Lyrical poetry, for example, appeared as a rather meaningless stylization of animal cries; and arts, culture, and human values were out of the scope of scientific philosophy. Science modelled after the paragon of physics appeared as the only legitimate mode of cognition, thus excluding wide realms of human culture from the philosophical system.

In a parallel way, the notions of "symbol" and "symbolic activities" remained unknown in American psychology, dominated as it was by behavioristic theory. In its "zoomorphic" orientation it was unable to see symbolism as easily the most important characteristic of human, as contrasted with animal, psychology and behavior. This is shown by the fact that the term "symbol" does not even appear in the index of leading textbooks of psychology nor is it found in Skinner's most recent presentation of psychological philosophy (1971). This is true of the mainstream of American academic and applied psychology ("behavioral engineering" by mass media, advertising, politics, etc.).

It is also true that, in contrast to American behaviorism, Russian psychology of the Pavlovian school (under the term of "secondary signal system"), some non-orthodox American psychologies (e. g., Werner, Maslow, Piaget, Charlotte Bühler), as well as psychiatry (e.g., Goldstein) early recognized the central role of symbolic activities in human psychology. Symbolism was, of course, a cornerstone of Freudian doctrine; but Freudian symbolism was almost exclusively concerned with one manifestation, i.e., the primary process with sex as the foremost "biological drive." Freudian symbols (e.g., phallic) would, in the present writer's opinion, probably be better defined as "pre-symbols." Higher forms of symbolism were insufficiently covered in Freud's theory of sublimation.

Possibly the most important development of psychoanalytic theory is Arieti's construct of the "intrapsychic self" (1967). In contrast to conventional psychoanalysis, it stresses the importance of cognitive and especially symbolic processes in normal psychology and psychopathology and distinguishes the primary (instinct, drives), the secondary (cognition, symbolic activities), and tertiary (creative) level of psychological functioning, with corresponding disturbances in psychopathology. Frankl's

(1959a, 1969) distinction of somatogenic, psychogenic, and noogenic neuroses is related to, though not identical with, Arieti's model.

The recent re-awakening of the interest in symbolism may perhaps be attributed to three main sources:

1. Cassirer's monumental work on symbolic forms appeared in German already in the 1920s. It would seem, however, that the sheer bulk and elaborate scholastic apparatus of this great work acted as a sort of deterrent and was prohibitive for its exerting a broader influence in psychology, general anthropology, philosophy, and so forth, for all of which it carried seminal ideals and facts. In retrospect, it would appear that only the American translation of this (1953-1957) and other fundamental work by Cassirer (1963) gained the broad influence it deserved, particularly by way of Wernerian psychology (Werner and Kaplan, 1963).

2. Susanne Langer's work (1942) was essentially based on Cassirer but developed the theory further in a thoroughly original way, actually presenting "philosophy in a new key" contrasted with the dominating positivism.

3. Unaware of these developments, some biologists, such as the neurologist Herrick (1956) and the present author, became concerned with symbolic activities. While Cassirer (and indirectly Langer) originated from neo-Kantian philosophy, the present author asked the biologist's question, What is characteristic of human as contrasted to animal behavior? and found the answer in the notion of symbolism (See Chapter 1).

Some Complementary Definitions of Symbolism

Cassirer: The Philosophic Viewpoint

Ernest Cassirer's *The Philosophy of Symbolic Forms* (1953-1957) is fundamental in the field. As indicated in the title, Cassirer's work concerns the *forms* rather than the contents of symbolism, thus covering a much vaster field and implying profounder problems than those envisaged in the present study. It is probably this *embarras de richesse*, evaluation of which would require hordes of specialists, together with the rather forbidding style of German academic philosophy, which has impeded the influence of Cassirer's work. The basic ideas are simple as well as revolutionary, but they must be laboriously extracted from an enormous matrix of history of philosophy, linguistics, mythology, epistemology, history and philosophy of science, and so forth, the factual details of which could be judged only by the specialist. Cassirer's shorter *Essay on Man* (1944) does no justice to the depth and breadth of the original investigation. Thus his work has not met with the understanding and application it deserves; otherwise, a superficial positivistic philosophy of science would not occupy the dominating position it still enjoys.

The range of discussion here is extremely small indeed in comparison to the vast background of Cassirer's work; nevertheless, it is interesting to note that from different starting points similar conclusions were reached.[1] Confirmation by independent experiment is the strongest verification of scientific theory; in philosophical discourse, independent development of ideas is what comes nearest to scientific "proof."

Symbolic forms, according to Cassirer, are essentially what Kant termed "categories," with, however, a radical revision and expansion which generalizes Kant's *Critique of Reason* into a Critique of Culture. The symbolic forms or categories are neither a passive framework of innate *a priori*s as Kant assumed, nor are they acquired by simple repetition as Hume and positivism contended; they are *creative functions* of the individual mind and culture concerned. Moreover, Kant's narrow system of "forms of intuition" and "categories of understanding" is exploded. The symbolic forms comprise not only those of "reason," that is, everyday and scientific cognition, but all activities characteristic of the human mind and culture, including language, myth, art, and so on. They are not simply "given" for every human mind or mind in general, but develop in close interaction with the several fields of cultural activity. Cassirer demonstrates how the categories of space, time, number, ego, existence, and so on, slowly emerge in interdependence with language, myth, and science. "Each language follows a *different* method in building up its system of categories" (Cassirer, 1953–1957, Vol. 1, p. 271). This is docu-

Etymology of Symbolism 61

mented by material of astonishing scope.

I believe that the roots of categories are both biological and cultural (Bertalanffy, 1955). A future "Table of Categories" would, therefore, have to be much richer than Kant's. It would have to take into account at least three major items:

1. Kant's Table (the effects of which are still present in modern positivism).[2]
2. Not only everyday experience and science, but all cultural activities have to be explored in their "symbolic forms." Discursive thought is but one of them; art, myth, language, poetry, mores, morals, law, and religion are others.
3. Those categories that are in part culture-bound. The Western way of thought is not the only possible one; other cultures may look at the universe (the I and the world) through quite different "spectacles." These spectacles largely depend on linguistic, cultural, and historical factors, and they deserve equal consideration.

Obviously, reformulation of the "symbolic forms" or categories implies a tremendous task which, in contradistinction to Kant's formulation, cannot be achieved by "transcendental" analysis of the Western world view and physics alone, but is subject to empirical research in a multitude of fields.

Here we are concerned with the more modest question of the definition of symbol. Cassirer did not make his definition quite explicit (see 1944, pp. 27–41) but it may be assumed that he would agree with the proposed criteria. This especially applies to the criterion of "free creation" which, as mentioned, has sometimes encountered criticism. Long after I had proposed it as being one of the criteria of symbolic, as compared to biological, activities, I found corresponding formulations in Cassirer's work (e.g., 1953–1957, Vol. 1 p. 92: "freely projected signs in language"; p. 141: "freedom which Adam possessed when he created the first name"; similarly in other places).

Royce: Correspondence in Signs and Symbols

Royce (1965) has emphasized "one-to-many correspondence" as a basic characteristic of symbols. This criterion is correct and is, in fact, one distinction vis-à-vis the signals of conditioned reaction. Indeed, a one-to-one correspondence applies to the latter: A certain tilt of the clock, an ellipse in contrast to a circle elicits the conditioned response, and the Pavlovian dog becomes confused or "neurotic" when one-to-one correspondence is impaired, for example, when the ellipse approaches a circle and the dog doesn't know whether to expect reward or punishment. In contrast, one-to-many correspondence is apparent in nondiscursive symbols, for example, when the same melody is used to accompany different situations or even different moods, for which well-known examples from musicology can be quoted.

However, Royce's contention is that discursive symbols ("signs" in Royce's terminology[3]) show one-to-one correspondence (for example, the unequivocal meaning of a mathematical formula), whereas (non-discursive) "symbols" provide for one-to-many correspondence (for example, the same dream may represent different entities; religious and artistic symbols imply a wealth of meanings).

Moreover, the "one-to-many correspondence" (and the freedom of combination or of "thought" it implies) characteristically distinguishes "symbols" from "signals" of conditioned reaction. Nevertheless, contrary to Royce's contention it appears that the one-to-many correspondence also applies to discursive symbols. Calculus or the Gaussian curve can be applied to entities of almost any sort; Newton's law applies to such different things as planets, apples, tides, and many others. As a matter of fact, this one-to-many correspondence is a prerequisite for the application of mathematics to science.

Furthermore, there is not only one-to-many correspondence, but also a many-to-one correspondence in symbols, both discursive and non-discursive. As Cassirer (1944, pp. 36ff.) has emphasized, it is a distinction between symbols (words, etc.) and signals (conditioned reaction) that the former are extremely variable and mobile. Different symbols (words) may represent the same entity within one language, and even more so within different languages; the same applies to representations by the spoken word, alphabetic writing, ideograms, etc. The same is true of non-discursive (for example, religious) symbols: The same saint is symbolized by liturgical texts, by images in paint-

Etymology of Symbolism

ing or sculpture, by hagiographic attributes, by churches bearing his name, and so forth. He could just as well be represented by Wagnerian *leitmotif* if church music had developed in that way.

This two-way street is attested to by the simple fact of homonymous and synonymous words in a language. This is not trivial and expresses an important trend in the evolution of symbolism, also mentioned by Cassirer. One trend is toward unambiguity and definiteness, as when vague circumlocutions in the vernacular are replaced by the clarity of mathematical formulation. But the opposite trend also is present, namely toward increasing mobility, one-to-many and many-to-one relationships. One-to-one correspondence is the primitive state; in primitive myth the name is an integral part of a god, and a prayer or rite must be performed according to a rigid formula in order to be efficient. Therefore, mythical and magical symbolism often puts primitive society into a straightjacket of actions resembling those of conditioning or of obsessive neurosis and obstructing change and progress. Similarly, in the development of speech—the child often has difficulty grasping that one and the same thing may have different names. Thus, one-to-one relations appear to be characteristic for primitive forms of symbolism in myth and language; slowly both universality (one-to-many) and plasticity (many-to-one) evolve. This appears to be an important characteristic of symbolism, a development which (considering the fixity of "schemata" in animal behavior) has hardly a parallel or precursor in the subhuman world. Thus, it appears that the relations between symbol and represented entity are highly complex and cannot be reduced to a simple formula.

Freud: The Psychoanalytic Viewpoint

One field of symbols (much spoken of) has not been included in the foregoing discussion. These are the symbols of psychoanalysis. The reason for this apparent lack is that the writer is wondering whether they should be termed "symbols" at all.

It is apparent that the symbols with which psychoanalysis deals do not correspond to the criteria proposed in this paper. In some vague sense symbols of sex and aggression can be called "representative." However, according to psychoanalytic doctrine, they are certainly not "freely chosen," but are characterized by their biologically determinate, or even obsessive, nature. They have, in Freud's version of psychoanalytic theory, nothing to do with tradition; the Jungian version, with its central concept of archetypes, would have to be strained to make symbols be "transmitted by tradition." Freudian symbols also are not "productive" in the sense defined above; they show no evolution and serve no purpose other than abreacting "psychic energy" in a devious way.

Since this is an important question, it will be well to reproduce some original statements in order to avoid possible misinterpretation. According to a well-known summary by Freud (1959, pp. 160ff.), "the number of things which are represented symbolically in dreams is not great."

> "All elongated objects, stick, tree-trunks, and umbrellas (on account of the stretching-up) . . . all elongated and sharp weapons, knives, daggers and pikes" and for different reasons "a nail file" are ♂. "Little cases, boxes, caskets, and stoves" are ♀. The dream of walking through a row of rooms is a brothel or harem dream. Staircases, ladders, and flights of stairs, or climbing on these, either upwards or downwards, are symbolic representations of the sexual act. . . . Tables, set tables, and boards are women . . . and "all complicated machines and apparatus in dreams are very probably genitals, in the description of which dream symbolism shows itself to be as tireless as the activity of wit. . . . Many landscapes in dreams, especially with bridges or with wooded mountains" are also symptoms of the same origin, and a Freudian disciple with the courage of his conviction (or is it a complex?) detects in those who love to wander in forests of erect trees a strong sexual proclivity. (Condensed by Jastrow, 1959, from Freud)

According to Kubie (1953), symbolism arises at three psychic levels: the realistic or conscious, the pre-conscious, and the deeper symbolic or unconscious level. In contrast to the original Freudian concept of strict representation, Kubie (and neo-Freudians in general) emphasize the one-many character of symbolism (as also remarked by Royce, 1965, pp. 15-23):

> Variations in the symbolic meanings of illness are among the subtlest and the most inconstant of all the variable forces which are at work in determining the ultimate psychosomatic picture. It should be unnecessary to mention this here; yet strangely enough, even in much analytic theory there is a tacit assumption that there is a one-to-one relationship between a concept and its symbol. This, of course, is never true on any level. There is not a simple relationship even between a table as real object, and concept table, or its verbal symbol. A table is a place to sit around, to put books and magazines on. A table is also a group of figures in a column. One tables a resolution. A table is a breast; a home; a mother; etc., etc. Since this is true on conscious and preconscious levels, inevitably it is true on unconscious levels of symbolic function as well. To imagine anything else would be curiously unrealistic, since we encounter the same multivalence of symbols in every dream that we analyze. A patient with a sore tongue dreamed of wanting a piece of cold ham or hot tongue. The hot tongue meant the tongue lashing she wanted to give her husband and analyst. It meant the tongue as a substitute for the genital intercourse of which her husband's impotence deprived her. It meant her old dependent infantile relationship to her mother. It meant the threat of punishment and death through the fantasy of a cancer of the tongue. Here it meant an identification with Freud, in a megalomaniac fulfillment of dreams of omniscience and omnipotence. In addition, there was the fact that the tongue actually was sore, which on the particular night played a part in the choice of the tongue to represent all of these struggles. (Kubie, 1953, pp. 74ff).

The present paper is not the place for a criticism of psychoanalysis or its therapeutic merits. Rather, we have to take the statements of psychoanalysis at face value, to analyze the alleged observations and theory, and to decide whether or not they fit into the concept of "symbolism."

The facts are that almost everything can elicit associations of a sexual nature. This is not unexpected, considering that sex is a factor pervading human behavior and the "dream interpretation" actually consists of two steps: the dream itself (which for that reason often is of sexual nature), and so-called free-association, which in fact is promoting sexual associations in the atmosphere of the psychoanalytic interview. In the wider sense of association psychology, almost everything can be associated with almost anything. In the narrower framework of psychoanalysis, almost anything can be interpreted as a sexual "symbol."

The underlying theory is, of course, that the censor or superego does not permit dreaming certain "forbidden" things, so the unconscious chooses the next best thing, i.e., replacing them by suitable associations or "symbols."

However, if there is a relation of, not one-to-many, but one-to-anything (the tongue, in Kubie's example, means fighting the husband, coitus, Doctor Freud, and innumerable other things) or anything-to-one (any "elongated object," according to Freud, is a penis "symbol"), we should not speak of symbol any more, but of free-playing "association."

Associations in the Freudian or a wider sense, apparently, are *material* for symbols, but not yet symbols, much as acoustic utterances (cries, babbling) are material for words but not yet words; or natural sounds (birds' song, tom-tom, etc.) are material for music but are not yet music. (Langer has an excellent discussion.) Perhaps the term "pre-symbol" would recommend itself for materials of a Freudian or similar nature. In order to become "true" symbols, they have to be consolidated—and this will invariably occur in the ways indicated by the three criteria in our definition.

In still other terms, association is obviously one basis of the symbolic process, but the symbol cannot be reduced to association as is presupposed when any object (dream, word, etc.) eliciting a sexual association is considered to be a sexual "symbol."

Two further remarks regarding the psychoanalytic theory of symbolism should be added. Symbolism has been called an "archaic" way of thinking by some psychoanalytic writers. It should be obvious that this contradicts all usual meaning of the word; following this usage, Beethoven's music, Michelangelo's statues, calculus, and nuclear physics would be "archaic"—and, supposedly, results of "repression" of a particularly wild "sexual instinct." This is *reductio ad absurdum*.

Etymology of Symbolism

Connected with this is the supposed role of repression in symbolic activities. For example, the development of dance as an art form has been considered as an attempt symbolically to satisfy frustrated sex drive (Jung after Hall and Lindzey, 1957, p. 101). The artificiality of such a construction should be obvious. Certainly, dance has to be considered much more as stylized play behavior, expression of *joie de vivre,* than as a result of "repressed" sex instinct, the latter of which primitive people would "uninhibitedly" satisfy by the real thing. One can see direct precursors of dance, at the instinctual level, in the often highly elaborate and "artful" mating ceremonies of animals, which precede sexual union in a strict pattern. Speaking here of "frustrated sex" makes no sense.

う

7
The Evolutionary Origins of Symbolism

The evolution of symbolism (rudimentary as our knowledge may be) warns us that the highest rational forms of symbolism (language, science) must not mislead us into assuming that creative symbolism is principally a process at the conscious or, in psychoanalytic language, at the "secondary" level. With respect to non-discursive and being-cognition symbolism, this "intuitive" origin needs no emphasis. However, in view of the empiricist attitude of the present, it cannot be repeated too often that creative intellectual work is also largely based upon "inspiration," that is, on "primary" processes in the unconscious and not on conscious data-collecting.

The process of "free creation" we used as a criterion of definition is largely on the preconscious level, not only in art, music, religion, but also in the "rational" constructions of science. Remember for example, Kekule's invention of the benzene ring (fundamental for organic chemistry) in a trance or dreamlike state (a sort of potentiated gestalt perception or fantasy); one will find similar intuitive insights in other pioneering achievements of science. Lorenz goes so far as to say that "Gestalt perception is capable of discovering unsuspected laws which the rational function of abstraction is totally unable to do."

> When little Adam invents his first noun, calling all dogs "bowwow," this is not owing to his first having abstracted the diagnosis of *Canis familiaris* as given by Linnaeus, but to the

functioning of his gestalt perception, which enables him to disentangle the essential configuration common to all dogs from the background of inessential differences and which permits him to perceive, in the aunt's peke, the neighbor's dachshund, and the butcher's mastiff, one common gestalt, that of the dog (Lorenz, 1960).

Subhuman Precursors: Ratiomorphic Behavior

Symbol forming is a "ratiomorphic" rather than a "rational," or reasoning, process—with the understanding that these expressions are not antithetical but are end points of a continuum. In an important study (1959) Lorenz has emphasized the *ratiomorphic* character of a great part of animal behavior. There are neurophysiological processes which, considering the complication of their function, equal the highest mathematical operations of our *rational* thought and which are, nevertheless, entirely devoid of any subjective phenomena. Amazing "computations" are performed by the mechanisms of perception, especially in constancy phenomena and gestalt perception. If, for example, Lorenz's small daughter (now a married lady) was able to pick out in the zoo the species belonging to the rather diversified bird family of Rallidae, this performance certainly approaches what taxonomists do by a highly involved symbolism of scientific language. Lorenz is certainly correct that gestalt perception is basic in the recognition of regularities ("laws") and, hence, in abstraction and, furthermore, in symbolism, language, etc., all of which (so to speak) attach a label to the phenomenal regularities so perceived.

Consciousness is a late offspring of the general ratiomorphic behavior of animal brains, being connected with the evolution of symbolic processes. The unconscious (biological) and conscious basically work on similar principles. In other terms, the brain is largely an "electronic computer" which, as such, lacks consciousness; only a small part of its calculations are "secondary processes" at the level of conscious, symbolic activities. Hence, there is no absolute gap between animal behavior and conscious, symbolic activity characteristic of man. One will

Evolutionary Origins of Symbolism

hardly err in assuming that ratiomorphic processes are a precursor of symbolic processes.

Modern computers show "ratiomorphic" behavior: They make calculations, have memory, may be goal-seeking, etc., and in general, behave in ways which in former times have been considered the privilege of rational and conscious minds. We have, however, no reason to assume that a machine that we have constructed has "consciousness." Exactly the same applies to the overwhelming majority of biological regulations: They are "ratiomorphic," but we empirically know by introspection that they are unconscious.

This being so, we are back at the epiphenomenalistic riddle: Why has consciousness evolved at all, if the job is done anyway by mechanisms lacking consciousness? As a rule, evolution does not produce characteristics that are useless; and to count consciousness among such "useless" characteristics as do occasionally occur implies that evolution of man was a particularly meaningless incident in evolution. The question looks different only when we give consciousness its due, if it is regarded, not as an inconsequential epiphenomenon, but as one which has functions not performed by ratiomorphically working mechanisms; that is, when man with his consciousness is the creator of a new world beyond physiological mechanisms and mere feelings of what is going on in the machine. Man as creator of his own universe—that of symbols and culture—justifies himself as a conscious being.

Biological Foundations of Human Behavior

We cannot answer the question of what was the "cause" of the evolution of man with his specific mental achievements. Nevertheless, some answer can be given to the more modest question: What were the predisposing factors making this evolution possible? No absolute gap between human and subhuman behavior is implied, although the uniqueness of man had to be defended. This, of course, is one aspect of the old problem of "levels" and "emergence."

Take an animal which (a) has social instincts, (b) has little specialized adaptation, (c) has a high degree of what we called "autonomous activity," expressing itself particularly in vocal utterances, and (d) does not have an elaborate innate or instinctual communication system. These characteristics will roughly correspond to those of the pre-human primates that were the ancestors of man:

1. Man certainly has innate social instincts. If he had not, there would not be a human society, which is made possible only by the fact that such social instincts are institutionalized. Hobbes's *bellum omnium conta omnes* hardly corresponds to the primeval human condition, because one can hardly imagine that society, matrimony, and other institutions would ever have evolved if social instincts had been lacking. On the other hand, human "togetherness" has limits; bees, ants and termites are much more socialized than man, their elaborate societies being built upon instinctive behavior.

2. Man's relative lack in specialized adaptations is well known; often it has been called the "biological helplessness" of man.

3. A certain creative power that can be ritualized well corresponds to the condition we find in anthropoids. Köhler (1925) relates how some of his chimpanzees invented "fashions" which, after a short while, were religiously followed by the whole group, being ritualized and institutionalized for a time. However, apes are lacking in particular vocal abilities, and Susanne Langer (1948, p. 85) has made the point that they don't talk because they don't babble as babies.

4. Finally, man does not have a highly elaborate innate communication system as have bees or ants; this, again, is connected with the relative poorness of his instinctual equipment.

It may be somewhat extraneous to the present discussion, but is of general interest, to put man into the larger perspective of the animal kingdom. If we look up the genealogical tree, we can distinguish two main branches of animal evolution. The first is called *Proterostomia* because of certain embryological characteristics. It leads from coelenterates to the various groups of worms, and culminates in the arthropods, and particularly the insects. As a matter of fact, insects represent one culminating branch of the animal kingdom. The other branch, called Deuterostomia, again broadly speaking, runs to *Amphioxus*, then to the series of vertebrate classes, and eventually culminates in mammals, including apes and humans.

In a rough approximation, one may say that these two phylogenetic branches also represent two possible approaches to higher forms of behavior. The one approach is by highly developed instincts, which finds its culmination in the societies of insects, such as bees and ants. The other possible development is that the instinctual basis of behavior remains relatively poor, and learning behavior is emphasized instead. This, in broad generality, applies to mammals and man. Incidentally, one may see why Lorenz's work is centered on birds: compared with mammals, birds have much more beautiful instinctual patterns—think, for example, of their elaborate mating ceremonies and dances.

If the four conditions mentioned are fulfilled, one might say that such a creature would be predisposed to develop a representative language based upon individual learning and symbols. If so, this of course would give this species, called man, a survival value and would be favored in evolution.

Conversely, we can roughly understand why apes do not talk—a question ably discussed by Langer. They are poor at socializing; a pack of apes apparently is less socialized than a pack of wolves or a flock of chickens. The ape is better adapted to its arboreal life than *Homo* is adapted to some particular environment. Apes do not babble as babies, and so do not have the spontaneous acoustical utterances out of which language can be formed. Finally, apes have nothing to say—they get along quite well with warning and expressive cries, and without a representative language symbolizing particular objects and features in their *Umwelt*.

Mind you, I certainly would not contend that what I have said indicates the necessary and sufficient causes leading to the development of language and symbolic systems in the species *Homo sapiens*. The foregoing only makes it somewhat more plausible that such evolution could have taken place. When primeval man developed the rudiments of language, this was a creative act which *a posteriori* can be understood as having had selective value. But it cannot be understood *a priori*, that is, we cannot deduce that under the circumstances just this development was necessary in order to preserve the species.

However, this is not a limitation peculiar to the problem of man's evolution; it equally applies to evolutionary problems in general. Evolutionary explanation can indicate necessary, but cannot indicate sufficient, conditions for a certain evolutionary change taking place. Consider, for example, the more trivial and familiar problems of morphologic evolution, say, the paradigm for the theory of natural selection—protective coloration. Certain moths amazingly resemble the pattern of tree trunks they rest on, and so are difficult to see; others mimic ill-tasting, and hence protected, species, and so are not eaten by birds. The explanation goes that protective coloration was useful for survival and eventually resulted in these marvels of imitation. So far so good. However, countless butterfly species which parade in the most conspicuous shapes and colors have nevertheless survived and flourished, presumably because they met the struggle for existence by other means—for example, overproduction of eggs and offspring. To quote another example: Ruminants have evolved a complicated, four-chambered stomach which is useful for digestion of the large quantities of vegetable food necessary for big beasts, and hence, this adaptation was accomplished in the evolutionary line of ungulates. The horse, however, equally big and herbivorous, does just as well with its simple stomach. Thus, even if one accepts the conventional theory of evolution in terms of accidental mutations, selective advantage, and survival of the fittest (i.e., organisms having the highest "differential reproduction"), it cannot be shown that any particular evolution took place by necessity, although *a posteriori* it can be interpreted as having had value for survival—otherwise, the species concerned would have been extinguished long ago.

This point is by no means trivial, although it is one hardly touched in textbooks of evolution. In admittedly metaphoric and anthropomorphic terms, one may say that evolution—even if the conventional explanation is accepted unreservedly—is a creative process, not explained by necessity of survival but productive of a multitude of phenomena from which necessity can choose. One might compare this with St. Augustine's doctrine

Evolutionary Origins of Symbolism

of "continuous creation," the contention that God did not create many species in the beginning (as Linnaeus has it), but rather manifested Himself as a creative force over the whole process of evolution. In this way modern concepts of evolution may be connected with ancient ones from the early Middle Ages.

The above also answers the sometimes discussed question why the disputable "wolf-children" (children that supposedly grew up without contact with other human beings) were incapable of language. It is only a paraphrase of what has been said that the invention of language, like any other great invention, was an act of creative genius. The average pre-human was just as unable to invent language as the average human was to invent calculus or relativity theory. Considering the small percentage of geniuses in any population, it is no wonder that the extremely rare wolf-children (if they have existed at all) did not belong to this minute group and were incapable of reinventing language.

Evolution of a "Universe"

Myth, magic, and the beginnings of human language, all closely interconnected, are doubtless near the origin of symbolic activities. Spoken language is unfortunately uninformative with respect to the origins of symbolism. There are many hypotheses as to its beginnings, but we do not know of any truly "primitive" language, that is, a language which would consist of representative signs (a "vocabulary"), but would not be propositional, would not have a "grammatical" structure. Even the languages of the most backward and savage peoples show no such primeval stage, but are no less (and, in some respects, rather more) complex than later and more civilized ones. Even the nonspecialist can follow the progressive simplification of language from Homeric to Attic Greek or from Chaucerian English to American.

Myth offers a more promising approach. It will be generally agreed that magical practices, verbal (the name of a thing gives dominance or power over it) or nonverbal (the paleolithic cave paintings as a means to ensure happy hunting), are close to the origin of symbolic representation. These, in turn, presuppose the mythical way of thinking. One would be tempted to propose as general categories of myth formation those of empathy (projection, personification, animism, anthropormorphism, etc.), of hypostatization (conceiving events in terms of persisting substances), and of reification (making concepts into "real" entities). This appears fairly to describe the universe of primitive man: He projects life and soul, similar to his own, into beasts, animals and things (empathy); inner experience is made into a substance (a half-material substance like mana, "blood" and "spirit" in primitive thought, still noticeable in Descartes's *res cogitans*); and words for things or mere activites become demons or deities (as is well exemplified by the deities in old Roman religion which personify the various activities in agriculture; this tendency still persists in concepts such as "force," "energy," and many others).

What psychology calls "empathy," and leaves as a very mysterious function, is basic for understanding of the "other ego" and hence for social intercourse; it is equally basic in mystical experience, magic, ritual, totemism, and art. On the other hand, hypostatization and reification of words and concepts is a very natural development. One must realize that *the invention of representative symbolism was the great discovery, the decisive event in the evolution of man*; no wonder it appeared to him as a sort of miracle and that the word became a higher, demonic, or even divine reality. Platonism, medieval realism of universals, reification of concepts (even in our scientific age) are more sophisticated forms of this primeval trend.

All this concerns processes at the preconscious or unconscious level; for this very reason it is refractory to satisfactory verbal formulation. Instead of speaking of a hardly understandable "projection" of the self into things, a better way to circumscribe the ineffable is perhaps to say that in primitive stages (which "modern" man has by no means outgrown) the differentiation between the *I*, the animate *thou*, the inanimate *it*, and the symbol (verbal or otherwise) is not yet fully developed. Using a psychiatric term, the ego barrier is still lacking; it is fluid or unconsolidated. The name or other symbol (pictorial, stone, etc.) not only represents but is the reality in question; man experiences himself not as an individual but as part of a collective "we" embracing his social group or even (in totemism and animism) a part of the whole of the world around him. The fact that this is a common stage in the evolution of humanity probably accounts for the rise of similar myths without cultural diffusion, as expressed in Jung's doctrine of archetypes.

In this sense, myth is neither a primitive philosophy or science (Tylor, 1874; Frazer, 1949) nor pre-logical confusion (Levy-Bruhl, 1910); rather it is a world view based upon categories that are essentially different from those of civilized man. That these categories are not outright impossible is attested by the fact that mankind got along with them fine for some hundred thousand years (much longer than the time of "high cultures," "common sense," and "science") and has survived. Overcoming the mythical categories is a comparatively recent event. It has given the dividend of ability to control nature; it has not been an unmixed blessing, in that it has created new problems while solving old ones.

Out of mythical thinking, of a floating experience encompassing the self and the world around, the representative function[1] gradually develops, with remnants of the primitive experience still remaining in modern man. All this, and many related facts, can perhaps be best expressed by saying that the I, the thou, and the it crystallize out from an undifferentiated flow of experience, and that this happens in individual development of awareness in the child as well as in the biological evolution of man and the cultural evolution toward individual consciousness.[2] It is not unilaterally caused by, but stands in mutual interaction with, conceptualization and language. "The I grasps itself through its counterpart in verbal action, and only as this latter becomes more established and sharply defined, does the I truly find itself and understand its unique position"(Cassirer, 1953–1957, Vol. 1 p. 277).

It is the representative function which creates a "universe." Immediate experience, such as perception of things, feelings, acts of will, and so forth, is momentary—dominating consciousness at one moment and gone the next. The past, in animal experience, consists only of traces left from previous conditioning, which influence subsequent behavior. Only when symbolism arises does experience become an organized "universe." Only then do past and future exist in their symbolic images, thereby becoming manageable. The past becomes part of the organized universe; the future can be anticipated by way of its symbolic stand-ins, and so can determine actual behavior. Thus, symbolism makes for the consistency of the universe: "Was in schwankender Erscheinung schwebt, befestiget in dauernden Gedanken" (Goethe, *Faust*).

The Demons of Language

The concepts of empathy, hypostatization, and reification may be used as circumscription for mythical experience, which, because of its non-verbal or pre-verbal nature is at the limit of linguistic expression. However, it must be borne in mind that this is a conceptual, and hence artificial, construction of what is in fact a unitary intuition. It is immediately apparent that mythical experience is not a thing of the past or an idiosyncrasy of savage tribes, but is still active in "civilized" man—often to his great peril. Anthropomorphism still persists in daily life and in science. It is even experienced by the well-ordered and systematic mind when, say, a person gets angry when some document plays hide-and-seek, concealing itself purposely, it seems, from the unfortunate seeker—acting in fact like a malicious hobgoblin. The German philosopher and writer Vischer (1879) has written an amusing story around this *Tücke des Objekts*. The same phenomenon can be found even in science: It took physics centuries to overcome the anthropomorphic concept of "force"; "entelechies" and similar vitalistic entities in biology are rather remnants of the old mythical experience than conscious "personificative fictions."

Even more precarious is another heirloom of mythical thinking: reification of words and corresponding concepts and, often enough, of words without corresponding concepts. Science is full of entities that are reified concepts which are only slowly dethroned as metaphysical entities. The century-long struggle in physics from an anthropomorphic concept of "force" to an abstract notion expressing certain quantitative relationships is again an example. The hypostatization of "matter" as ultimate reality in the mechanistic world view is another. The concepts of "force" and "matter" at the same time show the connection of categories and metaphysical notions with linguistic determinants (see Whorf, 1952; Bertalanffy, 1955). "Disease entities" in medicine are paler descendants of what past centuries called the "genius" of disease. In less developed fields, such as psychopathology, verbal magic still prevails. There is little difference between the notion of a demon obsessing an individual and controllable only by exorcism, and the notion of an "urge to kill" similarly obsessing a teenager, driving him to murder, and allegedly exorcised by the psychoanalyst. "Government," "society," "community" still are metaphysical entities, whether verbal entities are revered and/or abhorred under the biblical names of Baal and Jehovah or the modern names of "communist conspiracy," "free world," "race," or whatever name is used, depending on the adherent's position on this or that side of curtains and political fences.

Equally obvious, the trend to reification is of a primeval nature and has been beneficial in attempts to control the environment. Humanity would hardly have developed the tremendous algorithm of language very seriously and literally, endowing words with magical power and metaphysical reality. The so-called ontological proof of the existence of God (i.e., that we have to assume Him for logical reasons—hence He must exist) was shown to be fallacious, but similar fallacies still exist in plenty, because popular thinking today is much more muddled than, and inferior to, the sagacity of scholastics of the Middle Ages.

In the extreme case, of course, reification amounts to schizophrenia, making the patient's thoughts and words into internal realities. But the borderline always remains precarious, particularly if there is mass schizophrenia and, naturally enough, one madman cannot see the madness in others because he shares it. This is, roughly speaking, the world scene of today. After the episodes of the Enlightenment and the Century of Progress, the old demons again come to the fore, possibly in connection with the Decline of the West and the return to primitive "second religiosity," or rather demonology, to use Spenglerian terms. The feeble Voice of Reason, theoretically and in somewhat sectarian fashion raised by the General Semanticists and less dogmatically by some reasoning beings, appears of little avail against the old demons who are changing their names but not their nature.

Mind and Body

More detailed analysis would show that the basic dualism between the self and the world slowly develops (see Chapter 8, p. 92): The antithesis between body and mind, matter and soul, object and subject—obvious and cogent as it appears to us—is far from being a primeval datum. It is, rather a late conceptualization characteristic of modern Western civilization. One may remember Jacob Burckhardt's famous dictum on the birth of the individual with the Renaissance.[3] The various philosophical formulations—Descartes's *res extensa* and *res cogitans*, materialism and idealism, Fichte's *Ich* that posits the *Nicht-Ich*, Hegel's Spirit, Schopenhauer's Will, the physical universe of critical realism, the sense data of the positivists, and so forth—are all different attempts at such conceptualization within the framework of our culture and language structure. However, quite different conceptualizations are possible and have actually occurred in other cultural frames of reference. It would be unwise to take the modern dualistic concept of matter and mind as "real" and final—the more so as we witness its decay in modern physics and psychology.

Culture-Boundedness and De-Anthropomorphization

As the above brief discussion indicated, and more detailed analysis (Bertalanffy, 1955) would demonstrate, symbolisms and categories are largely culture-bound, an intimate interaction existing between ways of thought and language. Hence, categories change in history, as was briefly noted in the present study in the comparison of mythical categories to those applied by civilized Western man.

The progress of discursive symbolism is in the deanthropomorphization and de-reification of words and concepts. Anthropomorphic components are progressively eliminated; the system of thought becomes a conceptual construct, not identical with, but representing certain traits of reality. This concerns, first, "common sense" in practical life. Life in civilized and mechanized society would not be feasible if it were encumbered by mythical, anthropomorphic, and reifying thinking (although, as mentioned above, many remnants still remain). On the whole, the thinking of "practical man" must be realistic, applying categories that bring reasonable order into his universe and exclude myth and magic.

The ways of popular thinking and of science subsequently are accepted as categories which are *a priori*, that is, innate and cogent for any reasoning being. This, in a nutshell, is Kant's philosophy. However, science transcends these limits and proceeds to further deanthropomorphization, in two main ways. Physics notices that the categories and conceptualizations based upon everyday experience—such as Euclidean space, Newtonian time, and determined one-way causality—serve well enough in the world of middle-size dimensions to which *Homo sapiens* is biologically adapted. Hence the success of Newtonian physics and cognate fields of science. However, when science penetrates into the worlds of extremely small and extremely large dimensions—atoms on the one hand, galaxies on the other—these categories do not fit anymore. Therefore, even more generalized and de-anthropomorphized categories become necessary, such as the space-time continuum in the theory of relativity (the time coordinate having peculiar characteristics) and statistical causality in quantum physics. Thus, the construct of reality progressively loses all resemblance to the world of everyday life and becomes increasingly "unvisualizable," that is, different from and in contrast to the categories of everyday experience and practice. In other terms, the evolution of physics essentially shows that the seemingly innate and *a priori* categories of perception and thought in fact are conditioned by biological factors—being well suited for the *Umwelt* in which human beings usually move, but insufficient to deal with reality at large. Eventually, in modern physics, little more is left than a most general and abstract category of "order" expressed in mathematical formalism (Bavink, 1944). It is probable that future theoretical development in biology will follow a similar trend. Biology has to use categories alien to Newtonian physics, such as "wholeness," "organization," "directiveness," and so on, which are tentatively formulated in constructs of cybernetics, information, general system, and decision theories; and the overriding concept presumably will be an enriched and general category of "order."

In a different way, another group of disciplines (cultural anthropology, linguistics, history) leads to a similar result. These show that the popular categories of Western thinking and science present only one possible way of conceptualization. Other conceptualizations are possible within different cultural and linguistic frameworks. Even within Western civilization, categories have changed from those of mythical thought to the Newtonian-Kantian system to that of modern physics. Thus, categories are relativistic and conditioned by both biological and historical predispositions.

In such a synoptic view, which in many respects is related to that of Cassirer, myth and the origins of language mark the beginnings of differentiation between subject and object, reality and symbol. With progressive evolution this differentiation becomes increasingly established, so that the representative function of symbols is achieved in language, art, science, and so on. This, then in various "modes" covers the whole universe, objective and subjective. All elaborate symbolism corresponds to such differentiation, even though it is not universal for all rational beings or even the species *Homo sapiens* as a whole, but can be (and actually was and is) achieved differently in different cultures. In the highest exaltations of "peak experience" these differentiations, categories, or "rubrics" again disappear, so that experience becomes ineffable and can only be hinted at or allegorically circumscribed by existential symbols.

Summary and Conclusion

Subhuman precursors of symbolism were discussed, but they re-emphasize the uniqueness of symbolism in man. The evolution of symbolic worlds is identical with the creation of a human "universe," in contrast to the Uexküllian "ambient" (*Umwelt*) of animals, which is predetermined by their innate, anatomico-functional organization. Although it is impossible to explain the evolution of symbolism in man by necessary and sufficient causes, certain biological factors can be indicated which made this unique evolution possible.

An attempt was made to retrace the evolution of human universes, especially by taking myth as a primitive form which appears to permit easier investigation than the unknown origins of human language. Empathy, hypostatization, and reification were tentatively proposed as general categories of mythical experience. Considering, however, the processes at the unconscious or preconscious level, perhaps a more correct formulation is that the ego barrier is not primarily "given," but rather develops in phylogenetic evolution, individual development and cultural history. The consequences of this concept for the mind-body problem were briefly noted.

Considered as a philosophy, the concepts developed lead to a "perspectivist" view which is distinguished from both absolutism and relativism of human knowledge. Human knowledge depends on symbolic universes, which are based upon categories arising generally within the biological species *Homo sapiens* and within a given cultural context. In contrast, however, to complete relativism, human knowledge must represent certain traits or perspectives of reality, because otherwise it would be biologically fatal.[4]

The present writer arrived at this concept as a biologist, although he was influenced by historians of culture of the relativist school, such as Riegl, Spengler, Worringer, and others:

> As opposed to physicalism, we arrive at a standpoint which may be called perspectivism. Science, as well as everyday reasoning, consists of an array of conceptual schemes that help us to find our ways in the world. Any symbolic system that we apply represents a certain facet or aspect of reality. None of the world of symbols, the sum total of which is called human culture, is a full presentation of reality. However, *ex omnibus partibus relucet totum*, to use Cusa's expression: Each such aspect has,

though only relative, truth. (Bertalanffy, 1953)

This closely corresponds with the views of Dilthey in philosophy of history:

> Every world-view is conditioned historically and therefore limited and relative. Each expresses within its limitations one aspect of the universe. In this respect each is true. Each, however is one-sided. To complete all the aspects in their totality is denied to us. We see the pure light of truth only in various broken rays. (Dilthey, 1959; condensed)

The parallelism of the conclusions reached from science and from history is a strong argument in their favor.

8
The Mind-Body Problem: A New View

Up to the present time, the problem of body and mind belonged to the domain of philosophy. That is, it was discussed in terms of epistemology or metaphysics. An attempt is made here to approach the relations of body and mind as an empirical problem, utilizing the knowledge of modern biology: developmental, introspective, and social psychologies; psychopathology and psychiatry; cultural anthropology; linguistics; and the history of culture.[1] This consideration leads to the conclusion that the traditional Cartesian dualism cannot be maintained either with respect to immediate experience or to the constructs of physical science (neurophysiology) and psychology.

The Mind-Body Problem in Psychiatry

I begin this paper with considerable apprehension. The mind-body problem is one of the oldest problems in philosophy, tossed around for centuries and disputed by the most illustrious minds without finding a solution. Yet, I have further promised to discuss the problem in a "new view." What is there "new" that can be presented here—especially when I cannot promise any novel facts?

Thus, I am well aware of the difficulties of the task, and if I have chosen this topic, it must have been for cogent reasons. Up till now the problem was in the domain of philosophy. Even discussions of recent years, such as an extensive symposium published by Hook (1960) or the detailed dissertation by Feigl (1958), to mention but two, follow the well-trodden path and result in little more than a reshuffling of old ideas. A novel approach appears to be overdue, owing to the fact that the problem, far from being academic, is of immediate concern to the modern psychologist and psychiatrist. Three reasons come immediately to mind as to why this is so.

The first is the *high incidence of psychosomatic disorders.* Psychosomatic diseases are, of course, no new dicovery; Hippocrates knew about them long ago. However, there is suprisingly no prevalent discussion in present-day medical forums that they are indeed a medical problem of the first order; despite the fact that, since Hippocrates' time, there presumably has been a numerical increase due to the stress of modern life (or is it that we have only become more conscious of the problem?). "Psychosomatic," however, means nothing else than the mind-body problem expressed in medical terms. That business worries may cause ulcers is a clinical fact just as viruses cause pneumonia; but we have no simple way to reduce this to somatic medicine.

A second point is what may be called the *methodological helplessness of psychiatry.* We know only too well the limitations of modern medicine—from the cancer problem to trivial complaints like arthritis. But there is no other medical specialty which is so insecure in its therapeutic approach as psychiatry. Within a lifetime or less, we have seen the coming and going of radically different approaches, from psychosurgery to electroshock to the soft talk of the psychotherapist, drug therapy, and other measures. It is hardly an exaggeration to say that some therapy had a rationale but not much success, as is the case of lobotomy; while others (electroconvulsive therapy [ECT] and to a large extent psychotherapeutic drugs) were moderately successful but have no rationale and are purely trial-and-error. This state of affairs is not only unsatisfactory but implies considerable danger. It is true not only in a trivial way as with respect to fractures which may occur in ECT; even more alarming are such cases as reserpine depression—a side effect of the purely empirical administration of the drug which, before being recognized, led to therapeutic failures up to suicide. That psychiatry wavers between such extremes as drilling a hole in the skull and injuring the most important system in the body, and mere pep talk in some forms of psychotherapy, is an expression of the deep-seated insecurity in this field. This is not a question of ignorance which could be overcome by some new discovery of tomorrow—say, some new insight into the role of RNA (ribonucleic acid) in memory or some reformulation of the theory of instincts; rather it looks as if our ways of thinking, our basic concepts and categories, are inadequate. Obviously, a reconsideration of fundamentals is imperative—and an important part of this is the old problem of body and mind.

A third consideration may be called the *Erewhon* problem in modern society. I am not thinking of the more trivial aspect of Butler's utopia: that machines, instead of being servants to man, tend to become his masters. Rather, there is a more sophisticated part of the story. As you will remember, in *Erewhon* organic disease is a punishable offense that must be concealed at all costs. In contrast, social and moral dysfunction, such as a "mild case of embezzlement," are respectable and to be cured by practitioners called "straighteners." Isn't this largely what happens in our society? Being physically ill or merely getting old is a sort of punishable behavior; you will do well not to speak about it lest you be fired from your job. In contrast, mental disturbance and moral offense tend to become respectable and are referred to the psychiatrist. From murder to divorce and failure in school, we are inclined to consider all unsocial or ineffective behavior not in relation to a badly shaken value system or to problems arising in a complex society, but as interesting psychiatric cases, preferably the result of wrong toilet training or brain mechanisms gone astray.

I submit, and shall try to show in the following, that those and other inconsistencies and paradoxes in modern psychiatry are largely due to the fact that psychological theory is determined by an obsolete dualism of body and mind.

The Cartesian Dualism

As mentioned, the problem of body and mind was up till now in the domain of philosophy, a playground for more or less skillful conceptual acrobatics. However, the development of modern science tells a different story which is important for our purposes. What used to be philosophical problems of epistemology and metaphysics have become empirical questions to be investigated by scientific methods. This is true of the fundamental concepts in physics (space, time, matter, causality); and, to an extent, is also true for biology which nowadays explores problems such as wholeness, teleology, goal-directedness, and the like, which not long ago were the domain of a vitalistic demonology.

I believe a similar re-examination is due also for the problem which now concerns us. We have to apply whatever the fields of science (such as biology, psychology, psychiatry, cultural anthropology, comparative linguistics and others) are able to provide. In this way we shall not arrive at "final solutions" dear to the heart of philosophers, but we may make some progress relevant to psychological theory and psychiatric practice.

In order to start the discussion, it will be well to present the problem in its traditional form, even at the risk of being trivial. There are two connected problems which we may call the matter-mind and the brain-consciousness relations.

In our direct experience, we find two different sorts of things which we call material and mental. Material things are stones, trees, water, air, animals, plants, etc. To the mental realm belong sensations, emotions, illusions, thoughts, desires, motivations, etc. Experiences of material things are called "public"; that is, every suitably located observer will have a similar experience of physical things such as chairs, houses, rivers, and the rest. In contrast, mental experience is called "private"; the toothache I may happen to have is not shared by the dentist or anybody else, but only I am aware of it. The same is true of other mental experience.

We know further that all awareness in some way is dependent on states of our body and especially the brain. Thus the problem of the material and mental realms changes into the problem of the relations between brain and consciousness. Note that the viewpoint is essentially shifted. The distinction between material things and mental events is one of everyday observations. In contrast, the relations between brain and mind are highly sophisticated; when we speak of brain processes, we imply all relevant knowledge of physics, chemistry, neurophysiology, etc. That is, the universe of which science is speaking—molecules, chemical reactions, electrical currents, and what not—is not a content of direct experience, only connected with it by more or less elaborate chains of reasoning.

The problem is briefly this: There is matter somewhere in space. Matter sends out certain physical effects, say, electromagnetic waves. These eventually reach a physiochemical system of fantastic complexity, my body with its sense organs and brain. Light causes chemical reactions in the retina; they are propagated through the optical nerve and eventually arrive at the visual cortex. Now, as something fundamentally different from physiochemical processes, the sensation of red or green flares up. Conversely, there are mental events such as emotion, volition, motivation. These mental events are mysteriously transformed into physiological processes in the motor area of the brain. These are propagated through the pyramidal tract and eventually reach the muscles; a voluntary action takes place.

This is the dualism as it was first stated by Descartes in the seventeenth century. Descartes distinguished the *res extensa*, matter extended in space, and *res cogitans*, the conscious mind. We may apply somewhat different words, but the Cartesian dualism essentially remains the same.

Do not say that the Cartesian dualism is a dead horse or a straw man erected to be knocked down, as nowadays we have "unitary concepts" and conceive of man as a "psychophysical wholeness." These are nice ways of speaking, but as a matter of fact, the Cartesian dualism is still with us and is at the basis of our thinking in neurophysiology, psychology, psychiatry, and related fields.

I shall not enter into any detailed discussion of the traditional conceptions on the relations of body and mind and their shortcomings. As is well known, the main traditional answers show up in the theories of psychophysical parallelism, of interaction and of identity.

According to the doctrine of *psychophysical parallelism*, the chains of physical and of mental events run side by side, in some way corresponding to each other, but without mutual interference. But then the series of physical events is self-contained; it is fully determined by the laws of neurophysiology and ultimately of physics. What actually happens would occur in exactly the same way if mental events were absent; hence mind appears as an unnecessary and inefficient epiphenomenon in the physical world. But ideas do move matter—in the individual, society, and history. Observation, both introspective and behavioral, appears to show that behavior is determined by symbols, values, intentions, anticipation of the future—and these are something radically different from neurophysiological events, electric potentials, chemical reactions, and physiochemical processes in general.

The *theory of interaction* postulates an interference of mental in physical events and vice versa. This certainly corresponds to the unsophisticated impression, but remains unintelligible. How can an entity which, by definition, is non-physical interact with physical and chemical processes? This contradicts the very principles of physiology and physics. Conversely, physical processes should always lead to other physical processes; it is baffling, then, how some of them would produce something different in principle, namely, sensations, feelings, etc.

Finally, the *theory of identity* assumes that it is some ultimate reality that appears under the different aspects of physical and mental experience. But what, then, is this reality? The only way to conceive of it is along the lines of the only reality immediately experienced by us—that is, the specimen of our own consciousness, self, or psyche. But then we come to a panpsychism which is hardly acceptable and does not conform to fact. Out of the vastness of physical processes taking place in the universe, only a tiny slice have their mental counterparts—namely, those occurring in a living brain; and again, of the multitude of behavioral responses and of physiological events taking place in the brain, only an extremely small fraction is accompanied by consciousness. We have no indication of any difference between brain-physiological processes (i.e., electric potentials, synaptic phenomena, chemical and hormonal transmission of impulses, etc.) that are accompanied by conscious events and those that are not.

All these theories, including recent discussions of the problem, take the Cartesian dualism for granted. We, however, shall come to an essential revision if we follow our program and try to apply the testimony provided by various branches of modern science.

We may anticipate the result in a few sentences. The Cartesian dualism between material things and conscious ego is not a primordial or elementary datum, but is the result of a long evolution and development in the history of ideas. Other sorts of awareness exist which cannot be simply dismissed as illusory. Likewise, the dualism between material brain and immaterial mind is a conceptualization that has historically developed and is not the only one possible or necessarily the best one. As a matter of fact, the classic conceptualization of matter and mind, *res extensa* and *res cogitans*, no longer corresponds to available knowledge. We should not discuss the mind-body problem in terms of seventeenth century physics, but must reconsider it in the light of contemporary physics, biology, and behavioral and other sciences.

These are revolutionary or even paradoxical statements. Let us try to substantiate them.

The Testimony of Biology

We may start with the testimony of biology. To avoid introducing another Cartesian dualism—namely that of animals as soulless machines and only man endowed with a soul—we shall have to concede that a monkey or dog sees, feels pain, has certain desires and antipathies. If we do, the descent of the evolutionary ladder to ever more simple types of animals, nervous systems, and behavior, gives no indication of a break where precisely psyche leaves off and only reflexes and neurophysiological events remain.

On the other hand, we have good reason to believe that the universes apprehended by subhuman beings are very different from ours. To an extent, we are able to reconstruct them. This is the domain of von Uexküll's (1929, 1934) classic and colorful descriptions of the *Umwelt* or ambient world experienced by a dog, a fly, a starfish, a tick, a paramecium, and other animals.

Without going into details, two fundamental principles of ethology should be emphasized: The *Umwelt* of a given species is determined by the latter's organization—in particular, the structure of its receptor and effector organs; and the human *Umwelt*, i.e., the world as *we* experience it, is only one of a countless number of alternative "universes."

So far as we are able to tell, it appears that a principle of differentiation obtains. That is, what to us are exterior objects on the one hand, and our conscious ego on the other, slowly differentiate or crystallize out of an originally undifferentiated singleness of exteroceptive and proprioceptive experience. It is easy to see that the universe of objects around us is especially connected with sensation at a distance, particularly vision. Two factors are involved in the separation of physical objects and individual consciousness: first, suitable receptor organs, as just mentioned; second, the higher level of human awareness, symbolic factors, language, and thought, entering into perception and helping to establish the two worlds of objects around us and of conscious ego.

The Testimony of Developmental Psychology

This object-subject differentiation becomes much clearer when we come to social and child psychology. For the sake of brevity, we may take both together although we should always keep in mind that the individual development of the child is not simply a recapitulation of the evolution of the human race.

Anthropology teaches us that peoples in other cultures have world outlooks and conceptualizations different from ours. According to developmental psychology, the dualism between external world and ego, self-evident as it may appear, is in fact the outcome of a long development (see the excellent discussion Merloo, 1956, pp. 196ff).

The most primitive stage apparently is one where a difference between outside world and ego is not yet experienced. A psychiatric term is very useful in this respect. This is the notion of *ego boundary*. As is well known, psychiatry speaks of the breaking down of ego boundary in schizophrenia, where the border line between objects outside and what is merely hallucinated—between the "public" and "private" worlds—becomes vague or disappears. This, indeed, is part of the definition of schizophrenia; but a similar state of indefinite ego barrier obtains in normal development. The baby does not yet distinguish between himself and things outside; only slowly does he learn to do so—mainly owing to the obstacles and hindrances imposed by outside objects upon his activities.

In the next stage, the ego barrier develops but is not fixed in the sense of inanimate things outside and a feeling and volitional ego inside. Rather this is the stage of animism: Outside things—not only humans, but animals, plants, and even inorganic objects—are endowed with emotions and volitions, benevolent or, more often, malevolent and similar to those of the experiencing individual—the child or primitive human. Remnants of this animistic experience are still present in the adult personality. To cite a trivial example, we get "mad" at some object for which we are searching and which seems to behave like a malevolent hobgoblin, intentionally hiding itself. The same still applies to rather sophisticated ways of scientific thought. The animistic view is still in force in Aristotelian science. As Aristotle has it, each thing seeks its "natural place" and is endowed with a psychoid entelechy.

Meanwhile, the specific human faculty of speech, and symbolic activities in general, became developed. Here we come to a magical phase, where the animistic experience still persists, but with an important addition: The human being has gained the power of language and other symbols. However, no clear distinction is yet made between the symbol and the thing designated. Hence, in some way the symbol (e.g., the name or other image) *is* the thing, and manipulation of the symbolic image—such as uttering the name of a thing with appropriate ceremony, or depicting the beasts to be hunted, and the like—gives power over the objects concerned. The savage, the infant, and the regressed neurotic have no end of rituals for exerting such magic control.

Only in the last stage is the neat separation of external reality, ego, and symbols fully achieved. The ego boundary is established. Parts of experience—one's own body and mental processes—can be controlled immediately, while another part—the world outside—is only amenable to direct and limited control, either by physical action or by a mental interpolation of symbolic processes. It has been said that inanimate matter is an invention of the physicists of the Renaissance. Even then, it took a long time to de-anthropomorphize physics. We need only remember the long struggle about the physical concept of "force" which was at first conceived as an anthropomorphic principle and only lately de-personalized into energy as a purely mathematical concept.

This development may be formulated in somewhat different terms, but I think we can agree that it is essentially correct and can serve as a basis for further discussion.

I believe that the basic question in the mind-body problem is whether we should take the world outlook of the Western adult for granted and dismiss all others as primitive superstition or whether we should probe the bases of both everyday experience and the universe of science. I believe we have good reasons for pursuing the latter course.

The Testimony of Introspective Psychology

Philosophers have always begun by taking the duality of the physical and mental worlds as an unquestionable datum. However, any amount of psychological evidence shows that things are not that simple. Perception of the most trivial objects (tables, chairs, houses, people) is not a mere sum of sensations, or "sense data" (as positivist philosophers are fond of describing); perception is comprised of *Gestalten* of sense data plus memory, concept formation, verbal and other symbolic elements, conditioned behavior, and many other factors.

Experiment shows that even in perception made intentionally simple in the laboratory, motivation and expected gratification modify what is perceived. Considering the amount of individual learning, conditioning, and motivation that enters perception, it is extremely hard to say what proportion of the world, as we see it, is actually "public" in the sense of the positivistic definition. Even illusions participate in normal perception. Essential prerequisites for experiencing of the world around us as a well-organized entity are the constancy phenomena of psychology: constancy of size, shape, color, etc., of perceived objects. But the constancy phenomena are based upon discrepancies between sensation and perception—that is, upon mechanisms which, in psychological experiment, appear as illusions. It is not by intuitive experience but only by more or less elaborate functions of testing that we can tell what "really" belongs to perceived objects and what is illusion and delusion.[2]

Furthermore, I do not find a simple antithesis between physical objects outside and myself inside, but all sorts of intergradations. In the visual and tactual fields, our experience is not one of perceptions or simple sensations—which latter are artifacts of the laboratory—but various shaped, colored, etc., objects. In auditory experience, it is already less clear what is outside or inside. Is the *Art of the Fugue* an object in space or does it belong to inner experience? Equally unclear are the distinction and spatial localization of olfactory and gustatorial sensation.

Likewise, in introspection, the ego boundary appears fluid. The experience of my ego is not that of an immaterial entity, but is the universe of experience (proprioceptive in the widest sense of the word) that reports about a certain "material" thing which in physical language is called my body—just as exteroceptive experience is the universe of experience reporting about "material" things around me. I experience my feeling ego not as pure mind, but as subvocal speech (tension of certain parts of my musculature, etc.), my willing ego not as pure will, but as certain sensations of "pulling myself together," etc. Take this proprioceptive experience away and no consciousness of myself is left, in the same way as outside things disappear when I close my eyes. I believe William James, one of the most acute introspectionists, was quite right in emphasizing this observation, but this does not imply uncritical acceptance of the so-called James-Lange Theory as a physiological hypothesis.

Here, too, are all intermediates between sharp spatial localization and indefinite feelings. Pain is experienced in a well-circumscribed area of my body, the tooth or finger, in much the same way as a chair or tree is localized in the outside space of visual experience. But feeling well or sick, elated or depressed, is experienced by my body as a whole, rather like hearing a sound which is localizable only with difficulty. At the end of the scale is seemingly pure mental experience when, for example, in solving an arithmetical problem we nearly forget ourselves, although some tension of certain muscles, subvocal speech, etc., still may be observed.

For this reason, it is not to be conceded that, as Kant has had it, experience of the outside world is spatial, and inner or ego experience only temporal. The complex of proprioceptive experience that constitutes my ego is localized in space just as is the universe of exteroceptive experience. It is less sharply localized, it is true, but then in outside experience there are all shades from definite localization (vision, touch) to increasingly indefinite ones (hearing, the chemical senses).

And, of course, in pathological states, the ego boundary becomes blurred or disappears. A few micrograms of LSD will suffice to produce this effect. The voices hallucinated by a schizophrenic and those heard in normal discourse have equal reality value for the individuals concerned. But even with those who claim to be more or less normal, the borders between conscious ego, the unconscious, the physiological body, and outside objects are not rigid.[3] Every neurotic shows that vegetative functions that are purely physiological in the "normal" are psychological in him and vice versa. Yoga practice shows that physiological functions, otherwise involuntary and so supposedly concerning the body only, can be brought under conscious control. Or take the example of the phantom limb, whose presence after amputation a patient may still feel and experience. Conversely, a tool or machine may become a part of the experienced ego, a sort of extension or expansion of it. A good driver feels with the whole automobile. A good microscopist feels not with the tips of his fingers, but rather with the screws of his instruments.

Categories of Experience

In order to build an experienced universe from sensations and perceptions, mental operations are needed which Kant has subsumed under the concept of *categories*. But, contrary to Kant's view, the categories of space, time, number, causality, ego, etc., are not given once and for all as *a priori* concepts valid for every rational being; they are the product of a long and tortuous development. They are preconditioned by biological organization. As Lorenz (1943) has emphasized, neither man nor any other living being would have long survived if its perceptions did not mirror—in whatever distorted way—those features of the universe upon which the life of the species depends. But this implies only some sort of isomorphism, not an exact replication of reality. So far as human beings are concerned, the categories of experience further crystallize out in close interaction with social and cultural factors. Within the present framework we have to refrain from a detailed analysis, which has been given elsewhere (Bertalanffy, 1955). To give just a hint in what direction these processes may be sought, however, we refer to Piaget's (1959) investigations of how categories are established in the mental development of the child by interaction of organizational and behavioral factors; and to Cassirer's work (1953-1957) on how categories develop in cultural evolution, as studied by comparison of primitive and civilized peoples. It further seems that the formation of categories interacts with linguistic factors: The structure of language is both a conditioning factor and an expression of how the universe is organized. Here the so-called Whorfian hypothesis (Whorf, 1952) regarding the relation of the experienced universe with the structure of language would deserve further discussion.[4]

What this all amounts to is that the mind-body problem needs a much more intensive, scientific study than has ever been undertaken. Before we can even discuss mind and body in terms of the Cartesian dualism, we have to study the history, prehistory, and biology of these concepts. Taking them for granted and then trying to find some logical trick to coordinate them is, to use a famous simile, like observing the visible part of an iceberg and forgetting the much larger mass of ice below the waves of the sea.

Empathy and the Problems of Other Minds

We have still to look at the universe of science, and at the possible consequences, both theoretical and practical, of what has been said. Before doing this, I will glance at a further problem, that called in psychology empathy, and, in philosophy, cognition of other minds.

In some mysterious way we know that fellow beings experience anger, pain, pleasure, that they are endowed with mental experiences similar to ours. The behavioristic explanation is well known. It is that other minds are approached by a process of inference: If I feel pain or another emotion, I make a face or show other behavioral symptoms. Hence if I see you making a face of the type concerned, I infer from these behavioral clues that you feel corresponding pain or other emotions and that the ability for such inferences is acquired by a learning process.

In my opinion, the phenomenon of empathy and the experience of other minds is not a complex inference and even less so something verbally taught by the human mother to her infant, as some behaviorists have hypothesized. Rather, it is something very primitive or primeval; and empathy in civilized man is a pale remnant of a faculty of intuitive understanding which was much more highly developed in primitive man and even in animals. As a matter of fact, a pet dog or a budgie (a kind of parrot) appear empathically to understand my humors and intentions, sometimes to a degree surpassing the empathic understanding by the human partner. And this is the more remarkable because facial anatomies and expressive movements are so extremely unlike. The dog knows whether he is wanted or not; the budgie knows the location of my mouth, which he may kiss, or my eyes, which he must not, and does not, peck. Where does this knowledge come from? It can hardly be an innate schema bred by selection in evolution; budgies in the South American forest and humans were widely separated until budgies, not long ago, were imported as pets. The faculty does not appear to be learned—the dog has no opportunity to make comparative studies of human and canine expressions. Even the scientist who is eager to exclude all metaphysics can hardly avoid the impression that empathy (and related phenomena of mass psychology[5]) are a remnant of a collective unconscious, out of which individualized egos grew, but which still persists in traces. Of course, "collective unconscious" is a reification, a hypostatization into substance of what actually is only a set of dynamic happenings; but so are "force," "energy," and other respectable concepts of science. Using such models, we must only be careful to keep this in mind and not to make such concepts into metaphysical (and, as is often the case, divine or demonic) entities.

The connections of art, morals, religion, etc., with empathy need no emphasis, although they should be discussed in detail. Art and poetry pre-suppose empathy, cognition of other minds, animistic experience, the *tat tvam asi*—whatever term you choose to label this sort of awareness. In a way, the world experience of the child and the primitive is carried over into the most exalted manifestations of culture. However, one has to be a dry-as-dust positivist to consider the worlds of the artist and poet as merely an archaic relic. One will rather say that there are other, and perhaps higher, forms of awareness than those of ordinary life and of science; that the world of science is only one perspective of reality, highly useful and successful in its way, but not the exclusive one. For example, the unitive knowledge of which mysticism speaks is a form of experience said to be beyond ego and world, mind and body. That it is a genuine experience is confirmed by the fact of the independent appearance of the same mystical experience among humans of different creeds, cultures, and times.

Maslow (1959, 1963) has, I believe, well characterized in scientific terms what he calls "being cognition" in contrast to ordinary cognition. As is well known, Maslow distinguishes between the normal "deficiency cognition," which was the sole one taken into consideration by traditional Western psychology—that is, experience oriented toward coping with reality by means of adaptive perception and within an accepted symbolic framework—and "being cognition" attained in the peak of love, or at the height of a mystical or esthetic experience. Peak experience is non-utilitarian; it transcends the boundary between ego and non-ego; it renounces "rubricizing," that is, bringing things into the framework of symbolic categories; and it is detached from personal goals and anxieties.

We shall do well to adopt a *perspectivistic viewpoint* (Bertalanffy, 1955). The world view of science is admirable so far as it goes, that is, as a way of conceptual and technological control of nature. However, it is only one perspective of reality. The perspective of the artist and, in the last resort, of the mystic, is another, and this also is justified pragmatically: not in the way of controlling the world by technical marvels, but in the way of self-realization of the human personality.

The Testimony of Physics

When we finally come to science, it is not new to say that it is not essentially different from ordinary experience. Rather it is an expansion, refinement, and further conceptualization of experience. Again it has to be said that our Western science —essentially oriented by theoretical physics—is not the only possible one. A leading mathematician (Godel, quoted by Oppenheimer, 1956) has stated that it was purely an historical accident that our mathematics has developed along quantitative lines. Other, non-metrical forms of mathematics and corresponding models are quite possible, and are, in fact, found in recent developments (e.g., game and decision theory; Bertalanffy, 1962). Historically, there have been very different forms of science, that is, theoretical conceptualizations and models of what is experienced (Bertalanffy, 1955; Spengler, 1923).

The Cartesian dualism, the antithesis between soulless matter outside and immaterial soul inside, arises as a conceptualization characteristic of a well-defined state of Western science—the *res extensa*, the famous billiard balls, the atoms, moving in space according to the laws of classical mechanics,[6] and *res cogitans*, the working of the rational mind, the philosopher meditating on a comfortable chair in his studio, playing with highly abstract symbols. It is not a primeval datum, but rather the last ougrowth and flowering of an immense development occurring in the maturation of the human individual, in the evolution of man from lower animals, and in the history of culture from savage tribes to the rationalist philosopher of the seventeenth century.

Turning to modern science, what is left of these entities? Modern physics has destroyed the concept of matter except as a manner of speaking. The ultimate components of physical reality are not small bodies any more, but rather dynamic events, of which we can only say that certain aspects of their behavior can be described by certain mathematical laws.

Present psychological theory cannot, of course, be compared in sophistication with physics. Nevertheless, the general trend in physical and psychological science is similar. Physics expands the range of the observable by inventing instruments—microscopes, electron microscopes, and Wilson chambers—and so discovers entities beyond unaided sensory experience: cells, molecules, atomic particles, and so forth. Similarly, psychology expands the range of the observable by inventing suitable techniques. Psychoanalysis, for example, uncovers a realm of the unconscious which is not observed in naive experience. In order to explain what is observed, both physics and psychology construct models and theoretical systems which greatly surpass immediate experience and are linked with the latter only by long chains of deductive reasoning. In this way, what is eventually left in physics is a conceptual system permitting a more-or-less exact description of relationships among entities which, in their ultimate being, remain unknown. Psychology does the same. Constructs like id, ego, superego, drives, repression, and all sorts of other psychological hypothetical constructs or models are invented to describe and bring into a rational system certain relationships in experience. What these entities "are," metaphysically, remains undefined, and (as Freud sometimes has done) they may just as well be represented by mere letter symbols.

Thus, in the world picture of modern science, no ultimate reality is claimed for the little billiard balls and an immaterial mind to play with or be affected by them or, to use more modern terms, to interfere in the gaps of microphysical causality as left by the Heisenberg relation. Rather there is a reality which in exteroceptive experience is observed as a world of things, and in proprioceptive experience as the ego. In science, this is described, with respect to certain structural aspects, by physical and psychological theories.

Culture and Values

Finally, the uncritical acceptance of the Cartesian dualism may have led us to forget that matter and mind by no means cover the entire field of reality. There are many realities which are neither physical nor mental, but which are beyond and outside the Cartesian antithesis. We cannot enter into a detailed discussion or definition, but obviously, beside responding to biological needs, human behavior is fundamentally determined by realities which, in a loose way, we may call cultural, symbolic, spiritual values, and the like. It is easy to see that they fall into neither of the Cartesian categories—they are neither physical, like rocks and animals, atoms and chemical reactions; nor are they mental, like feelings and thoughts, motivations and other psychological constructs. I suggest that if one is to think this through, he should start with trivial facts in our society—say, the Bureau of Internal Revenue as a very real entity which nevertheless is neither a physical thing nor, unfortunately, a mental hallucination—and go on up to the sublime achievements of culture called science, works of art, religious values, and so forth. One should think over whether a Beethoven symphony, a Rembrandt painting, or the system of physics can be defined in terms of the categories of "physical" and "mental." It will easily be found that they cannot be.[7] But it is just such realities on the higher or symbolic level which determine the most important part of human behavior (see Chapter 1).

Again, this is not metaphysical speculation, but reflects on psychiatry. I have said that hitherto psychiatry was determined by the Cartesian antithesis—being either physical (like psychosurgery, shock, drugs, etc.) or psychological (in terms of attempting to treat the individual mind). It is a consequence of this that it could not deal with broad fields of mental health and disturbance. Much is said, for example, of existential neurosis arising not from frustration of biological needs or from particular conflicts, but from the meaninglessness of life in modern society, a world in which values, purposes, and goals have collapsed. What has this to do with electrical potentials in the brain or unsatisfied drives? Nevertheless, the affliction may be sufficient to provoke suicide. Or, in the problem of delinquency (assuming, for the sake of argument, that it is a psychiatric problem, which could well be disputed), it has been said that a new style of crime has appeared—crime not because of need, for material gain, or out of passion, but crime for "thrills," or for establishing a reputation as a "tough guy." But what can be done about it with tranquilizing drugs or conventional psychotherapy? There is no disturbed brain physiology, nor do phenomena like existential anxiety and new-fashioned crime fit into the usual psychological or Freudian categories. The only thing that can be said is that these are disturbances which originate in a breakdown of the value system, loss of goals of life and of spiritual orientation—that is, they come from that third realm, other than matter and mind.

I believe I have succeeded in showing that our topic is not a purely academic one, a playground for abstract philosophies, but one intimately connected with great problems of our time and society—and especially with psychiatric questions.

Isomorphism and General Theory

To put our discussion of the mind-body problem to practical use I would like to suggest some new approaches in both theoretical and practical areas.

We have agreed, I presume, that physics and psychology (both taken in a wide sense) are conceptual constructs representing certain aspects of reductionism. The concepts of psychology cannot be reduced to those of neurophysiology (physics), as should have been clear from the start. Neither is the mental world an epiphenomenon to the physical world of atoms, chemical reactions, electric currents, etc., which is completely determined in itself by the laws of physics, so that the mental series would represent an inconsequential and unintelligible duplication. Both the worlds of physics and of psychology are constructs to bring certain aspects of the experienced universe under the rule of law.

Nevertheless, excluding reduction of psychology to neurophysiology, we can indicate what their relation is and how unification of both fields may be sought. In order to relate them we must postulate an *isomorphism* between the constructs of psychology and neurophysiology. This is both the minimum hypothesis required and the maximum hypothesis permitted by science: the minimum hypothesis, because neurophysiology would make no sense without correspondence to mental processes; the maximum hypothesis, because this is the most we can say without metaphysics.

However, one must be careful not to take this isomorphism in a simple and naive way. It does not imply any simple similarity between psychological and brain-physiological processes, say, between visual *gestalten* and corresponding electric fields in the brain. Here the simile of modern "thinking machines" is illustrative. For example, we can well imagine—and I think this would even be technologically feasible—a machine that builds automobiles in a fully automated way. This would imply a program running through a computer and a series of allied machines. But the program—perhaps in the form of a punched tape—has no apparent resemblance to the automobile produced, although, in the way of a code, it is isomorphic with the latter. Incidentally, something similar is actually the case in biology, namely with respect to the genetic code of protein synthesis contained in the nucleic acids of the chromosomes. If, as seems probable at present, the memory function is connected with the RNA of neurons, this also would presuppose a presently unknown manner of coding. Hence, isomorphism between psychological and neurophysiologic happenings need not presuppose any simple resemblance between both series.

Now, in what way can neurophysiological and psychological theory further be unified? I believe we can give quite a definite answer to this question also. As has been stated, the unification will not be along the lines of taking the constructs of physics as absolutes and reducing them to psychological constructs. Rather I see the unification of physiological and psychological theory in constructs which are generalized with respect to both, and in this sense are neutral with respect to physics and psychology. We have a fair idea what such generalized theory may look like (see Chapters 9 and 10). Recent theory construction in cybernetics, information theory, general system theory, game and decision theory, etc., elaborates constructs precisely of this kind—that is, constructs that are neither physical nor psychological, but are applicable to both fields. Admittedly, this is only a beginning, but I believe the problem is rather clearly posed: the formulation of a generalized theory within which both psychical and neurophysiological constructs appear as specifications.

Overcoming Physicalism

Such theory, even in the vague outline which we can give it at present, can have very definite practical consequences.

The dichotomy in psychiatric therapy between physical and psychological methods is a consequence of the philosophical antithesis of body and mind. The conventional model of brain function was physicalistic, that is, adopted from traditonal physics without regard to biology. This model conceives of the organism as an essentially inert system which is activated only by external factors. This is the way ordinary physical systems behave; from it follows what may be called the "automaton model" of the living and behaving organism. It can easily be shown that this automaton model predominated until now in psychology (Bertalanffy, 1962), in terms of the stimulus-response (S-R) and other schemes. The concept of psychological and social homeostasis; Freud's principle of stability according to which the tendency of the organism is to release tensions and come to rest in an equilibrium state; the consideration of mental illness as disturbance of such equilibrium, and the consequent ideal of man as a robot to be maintained in optimal psychological and social homeostasis and adjustment to given conditions—all this and much more are consequences or different expressions of the automaton model.

However, this model is unsatisfactory in theory and dangerous in its practical consequences. Modern biology teaches us that the organism is not an ordinary physical system, that is, one corresponding to conventional (or, rather, obsolete) physical theory. It is a so-called open system, and among the characteristics of open systems is that they not only respond to stimuli but show what may be called inner or autonomous activity (see Chapters 4, 9, and 10). This, of course, corresponds to the experience of both classical (Bertalanffy, 1952) and recent neurophysiology (Hebb, 1955). It will suffice here to mention arousal systems, such as the reticular activating system, to illustrate the importance of autonomous activity.

And, of course, what has been said also corresponds to experience in psychiatry and mental health. If, according to the S-R scheme, the supreme tendency of the psychophysical organism is to satisfy biological needs, why is it that, in our so-called affluent society where the biological needs of hunger and sex and daily life in general *are* satisfied as never before, we have an unprecedented increase in mental cases? I think that the fact that fifty per cent or so of hospital populations are psychiatric patients is the most dramatic illustration that something is fundamentally wrong with conventional principles. And I believe it is possible to define what is wrong rather clearly, and to draw the practical conclusions.

The overcoming of physicalism leads us to replace the S-R scheme and automaton model by a more realistic one which approaches the psychophysical organism as an internally active system. This implies a re-evaluation of both psychological theory and practice. As a matter of fact, with notions such as emphasis on activity and creativity, self-realization, and the like, psychology has already escaped the fetters of the S-R model and is tending toward new concepts of the kind I tried to indicate.

Summary and Conclusion

A modern reconsideration of the mind-body problem must consider, first, recent developments in biology, developmental psychology, cultural anthropology, linguistics, psychopathology, theoretical physics, etc. Second to be taken into account are developments in modern physics and biology that show that problems formerly considered to be epistemological, philosophical or metaphysical (e.g., those of space, time, causality, wholeness, directedness, etc.) have become increasingly subject to empirical research. The same will apply to the mind-body problem.

The problem encompasses two levels, namely, those of direct experience and of the concepts of science. In direct experience (introspection) the antithesis between ego and non-ego ("material things") is a result of a long developmental process, biological as to the evolution of man, psychological as regards child psychology, and cultural as to human history. It is not a self-evident category nor is it *a priori* for every human or rational being. Other forms of consciousness, such as peak experience in emotional climax, art, mysticism, etc., cannot be disregarded as mere primitive precursors of the so-called objective world view of the average Westerner in the twentieth century. They are different forms of cognition in their own right.

In science, the antithesis of "matter" and "mind" is a conceptualization characteristic of the mechanistic model and world view of physics. "Mind" and "matter" are reifying conceptualizations that become increasingly inadequate in modern science. The concept of "matter" in the classical sense is abandoned in modern physics. Similarly, the concept of "mind" is a reification of what actually is a dynamic process. This concept no longer holds in present science, as is shown, for example, by the concept of the unconscious, which does not fit into the Cartesian dualism.

The classical dualism omitted precisely that realm which is specific to human, as compared to animal behavior and psychology: the field of culture, symbols, values, etc., which are neither "physical" nor "mental," but have their own autonomous laws. Basic as well as clinical psychology must recognize this realm because it is precisely the sphere of specifically human behavior, and new developments in both fields may be expected from a proper acknowledgement of this fact.

Both physics, including neurophysiology, and psychology, including unconscious processes, are theoretical constructs aimed at explaining, predicting, and controlling observable events; they are connected with the latter only by extensive chains of reasoning. Both are progressively de-anthropomorphized, that is, the properties characteristic of human experience and *Umwelt* are progressively eliminated. What eventually remains are conceptual models and relations serving the purposes of explanation, prediction, and control. This process is far advanced in physics and beginning in psychology.

The new approach to the mind-body problem constitutes a working hypothesis that should lead to new insight, to useful consequences in both theoretical psychology and psychiatry. A unification of both conceptual systems—those of neurophysiology and of psychology—appears to be possible by use of models that are neutral and superordinated to both. The beginnings of such a development can already be noted.

It is a necessary postulate for neurophysiology and psychology that their constructs are in some way isomorphic. Such isomorphism need not involve similarity of neurophysiological and psychological events; the concept of coding gives an indication of isomorphism without any direct similarity or resemblance.

The new approach in the mind-body problem leads to a more realistic model of the psychophysical organism; it amends the conventional S-R and automaton model of neurophysiology and psychology.

9
General Theory of Systems: Application to Psychology

Definition

General system theory is intended to elaborate properties, principles and laws that are characteristic of "systems" in general, irrespective of their particular kind, the nature of their component elements, and the relations or "forces" between them. A "system" is defined as a complex of elements in interaction, these interactions being of an ordered (non-random) nature. Being concerned with formal characteristics of entities called systems, general system theory is interdisciplinary, that is, it can be employed for phenomena investigated in different traditional branches of scientific research. It is not limited to material systems but applies to any "whole" consisting of interacting "components." General system theory can be developed in various mathematical languages, in vernacular language, or can be computerized.

Reasons for the Present Interest in General System Theory

Psychology in the first half of the twentieth century was dominated by a general conception which may be epitomized as the *robot model* of human behavior. Notwithstanding the great differences between these theories (such as psychoanalysis, classical behaviorism and neobehaviorism, learning theories, computer models of the brain and behavior, etc.) they shared certain presuppositions. Among these was the concept of the psychophysical organism as being essentially *reactive*, that is, behavior is essentially to be considered as a response, innate or learned, to stimuli (the S-R scheme). Similarly, the same idea was basic in the consideration of psychological and behavioral phenomena as re-establishment of a disturbed equilibrium (homeostasis), as reduction of tensions arising from unsatisfied drives (Freud), as gratification of needs (Hull), as operant conditioning (Skinner), etc. The needs, drives, tensions, etc., in question were essentially *biological*, while the seemingly higher processes in man were considered secondary and eventually reducible to primary biological factors such as hunger, sex, and survival. For this reason machines, animals, infants and the mentally disturbed can provide adequate models for the study and explanation of human behavior and psychology: machines, because behavioral phenomena eventually are to be explained in terms of machine-like structures of the nervous system; animals, because of the identity of principles in animal and human behavior, and the better amenability of the first to experimental investigation; infants, because in these—as well as in pathological cases—the primary factors are better recognizable than in the normal adult. Scientific investigation was aimed at the detection of elementary entities (sense impressions, reflexes, conditioning processes, drives, factors, etc.) juxtaposition of which would provide explanation of complex behavior.

For a variety of reasons the robot model proved to be unsatisfactory. The trend toward a new orientation was expressed in many different ways, such as in the concepts of developmental psychology, genetic epistemology (Piaget), or progressive differentiation (Werner); various neo-Freudian developments (e.g., Rogers's client-centered therapy, Schachtel's emphasis on activity in cognitive processes); ego psychology; the so-called New Look in perception; self-realization (Goldstein, Maslow); personality theories (e.g., Murray, Allport); phenomenological and existentialist approaches; sociological concepts of man (Sorokin); and others. Although these developments were most diversified in intent and content, they seem to have one common denominator, namely, to take man not as a reactive automaton or robot but as an *active personality system*. If so, general system theory may provide a general conceptual framework.

Some System-Theoretical Concepts in Psychology

As is apparent from the above, the bearing of general system theory in psychology is not in the way of some startling new discovery. Rather it turned out that basic preconceptions or categories of psychological theory and research which were taken for granted in a positivistic-mechanistic-behavioristic approach, were found to be insufficient. This was part of a much broader reorientation. For essentially similar problems appeared in biology (the "organismic" conception); in the social sciences (question of supra-individual organizations); in applied fields (e.g., technology of man-machine systems as compared with physical technology); and even in physics (multi-variable interactions in "organized complexity" vis-à-vis linear Newtonian causality and unorganized complexity in statistical phenomena; Weaver, 1948). In any of these developments, problems circumscribed by notions such as wholeness, organization, goal-directedness, hierarchical order, regulation, etc., appeared as central; that is, characteristics which were not only bypassed, but *a priori* excluded in the classical mechanistic universe. The answer to this quest, a general "science of systems," is admittedly in its first beginnings at present.

The present review cannot give any systematic presentation. Only by way of example, some system-theoretical notions in their application to psychology will be indicated; for more detailed study the reader is referred to the work quoted.

Organism and Personality

In contrast to physical forces like gravity or electricity, the phenomena of life are found only in *individualized entities* called organisms. Any organism is a system, that is, a dynamic order of parts and processes standing in mutual interaction (Bertalanffy, 1928, 1952). Similarly, psychological phenomena are found only in individualized entities which in man are called personalities. "Whatever else personality may be, it has the properties of a system" (Allport, 1961).

The "molar" concept of the psychophysical organism as system contrasts with its conception as an aggregate of "molecular" units such as reflexes, sensations, brain centers, drives, reinforced responses, traits, factors, etc. Psychopathology clearly shows that mental dysfunction is a *system disturbance* rather than loss of single functions. Even in localized traumas (e.g., cortical lesions), the ensuing effect is impairment of the total action system, particularly with respect to higher and hence more demanding functions. Conversely, the system has considerable regulative capacities (Bethe, Lashley, Goldstein, etc.).

Closed and Open Systems

A living organism is an open system, i.e., a system maintained in import and export, building-up and breaking-down of material components; in contrast to the closed systems of conventional physics which do not exchange matter with environment. Some characteristics of open systems as contrasted with closed systems are the following:

1. Closed systems *must* (according to the second principle of thermodynamics) eventually reach a state of *equilibrium* where the system remains constant in time and (macroscopic) processes come to a stop. On the other hand, open systems *may* attain a *steady state* (certain conditions presupposed); in this the system also remains constant in time, but processes are going on and the system never comes to "rest."

2. The state of equilibrium eventually reached in closed systems is determined by the initial conditions. In contrast, if a time-independent, steady state is reached in an open system, this state is independent of the initial conditions and only depends on the system conditions (such as rates of transport and reactions, etc.). This property of open systems is called *equifinality* and accounts for many regulations in living systems. Whereas in the familiar closed systems of physics the final state is determined by the initial conditions, in open systems, as far as they attain a steady state, this state can be reached from different initial conditions and in different ways;

it is thus equifinal (Bertalanffy, 1950a). Many regulative processes take place in such way that the same final state or "goal" is reached from different initial conditions, in different ways, after indefinite disturbances. Because such behavior is impossible in closed systems and conventional machines, equifinality appeared to be a vitalistic feature, violating physical laws; but it is a necessary consequence of steady state in open systems.

3. Closed systems develop toward states of maximum entropy, that is, states of increasing probability and disorder. In contrast, in the living world—as in individual development and in evolution—a transition toward states of higher order ("anamorphosis") is found which seemingly contradicts physical law as expressed by the second principle of thermodynamics. The apparent contradiction disappears in the generalization of thermodynamics to include open systems (so-called irreversible thermodynamics). For in open systems, there is not only production of entropy due to irreversible processes, but also transport of entropy, for example, if matter with high free energy content is introduced into the system. The balance of entropy production and transport may then well be negative, that is, open systems may exhibit anti-entropic processes and develop toward states of higher order, differentiation and organization. Possibly a future integration of irreversible thermodynamics and information theory will permit deeper understanding of the fundamental problem of "anamorphosis."

The significance of the theory of open systems for psychology was discussed by Krech (1950). Here only a few consequences will be indicated. For example, neurophysiological *Gestalten* are to be considered as "open" rather than as "closed" systems. One of the basic principles of Gestalt theory is the *law of pregnance*, stating that perceived *Gestalten* tend towards the maximun regularity, symmetry, and simplicity possible under the circumstances. This is considered to be a consequence of the attainment of a steady state in the corresponding neurophysiological *Gestalten* (cf. 1. above). Krech also emphasizes that an open system model in neurophysiology "suggests that some experienced forms, under some circumstances, may tend toward *increased* heterogeneity and *increased* complexity" (1950, p. 353) (cf. 3. above).

The principle of equifinality (cf. 2. above) would correspond to the concept of "vicarious functioning" in behaviorism (Hunter, Boring, Brunswik, and others). It appears that brain centers are not machineries fixed from the beginning, but that they differentiate in a process of progressive mechanization. Thus a center is not a sharply circumscribed region; its functional potencies usually extend over larger parts of the central nervous system (C.N.S.). In the normal course of events that region which can do it best governs the function; it is the leading part. If this region is injured, other parts which have the same potency, though to a lesser extent, may do the job and thus give rise to vicarious functioning. Another part of vicarious functioning may be taken care of by feedback mechanisms. If, for example, the same goal can be reached by different ways of locomotion such as running, flying, swimming, etc., then, as in a guided missile possessing different kinds of locomotor apparatus, the effect may be due to a switching-over from one type of locomotion to another, each feedback-controlled, and thus lead to the effect mentioned. Finally, there may be phenomena of vicarious functioning which are due to true equifinality, that is, attainment of the same steady state from different initial conditions.

The Active Organism

Natural behavior encompasses innumerable activities beyond the S-R scheme, from exploring, play, and rituals in animals to economic, intellectual, aesthetic, and religious pursuits, to creativity and self-realization in man. The behavior of any organism in innumerable activities under natural circumstances goes far beyond reduction of tensions or gratification of needs (e.g., Allport, 1961; Berlyne, 1960; Piaget, 1959; Schachtel, 1959). All such behavior is performed for its own sake, deriving gratification (function pleasure, after C. Bühler) from the performance itself.

Even without external stimuli, the organism is not a passive but an intrinsically active system. Modern research has shown that autonomous activity of the nervous system, resting in the system itself, is to be considered primary. In evolution and development, reactive mechanisms (reflexes, feedback circuits) appear to be superimposed upon primitive, rhythmic-locomotor activities. A stimulus (i.e., a change in external conditions) does not *cause* a process in an otherwise inert system; it only modifies processes in an autonomously active system (Bertalanffy, 1937). Similarly, automatic sequences of impulses, the so-called hereditary coordinations, which are often discharged without even external stimuli, play a predominant role in instinctive behavior (Lorenz, 1935).

In contrast to machine theory, this primary activity is one of the essentials of organismic biology in general, and of the theory of the organism as an open system in particular. The organism appears as a flow of processes which can be considered, for certain purposes and in a first approximation, to be a steady state. Superimposed on the steady state are smaller process waves, a rhythmical storing and discharge of impulses, the type of relaxation oscillations which give rise to autonomous activities and to rhythmic-automatic functions in particular.

Autonomous activity is the most primitive form of behavior (e.g., Bertalanffy, 1952; Carmichael, 1954; Herrick, 1956; Holst, 1937; Werner, 1957). It is found in brain function (Hebb, 1949) and in psychological processes. The discovery of activating systems in the brain stem (Magoun, 1958) has emphasized this fact in recent years.

Homeostatic and Dynamic Regulation

Homeostatic and *dynamic* regulation are two allied but different models, applicable to different but sometimes also to the same phenomena. Homeostasis refers to regulations effectuated by feedback mechanisms, such as that shown in Figure 2. That is, in a system responding to outside disturbances, part of the output is monitored back to the input so as to control the

Figure 2 The Feedback Scheme

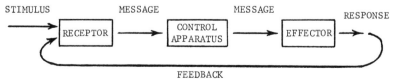

system's function, either to maintain a desired state or to guide the system toward a goal.

The feedback scheme is the basic element of classical cybernetics. However, within the framework outlined here, cybnernetics is concerned with an important but not all-inclusive subclass of systems. The differences between classical cybernetics and general systems can be seen by inspection of the above diagram. Namely, feedback regulation is by way of linear and unidirectional (although circular) causality, while regulation in general (and especially in open) systems is by way of *multivariable* interaction. Regulations of the first kind are based upon pre-established arrangements ("structures" in a broad sense), regulations of the second upon *dynamic* interaction. Feedback arrangements are systems closed with respect to energy and matter, although open to information; open systems, on the other hand, are better understood by recourse to generalized principles of kinetics and thermodynamics. As a rule, feedback circuits are superimposed on, and develop from, primary regulations as secondary regulative mechanisms.

From the above it follows that the concept of homeostasis—frequently applied in psychology—covers animal behavior only partly, and an essential portion of human behavior not at all. Its limitations have been aptly summarized by C. Bühler (1959):

> In the fundamental psychoanalytic model, there is only one basic tendency, that is toward *need gratification* or *tension reduction*. . . . Present-day biological theories emphasize the "spontaneity" of the organism's activity which is due to its built-in energy. The organism's autonomous functioning, its "drive to perform certain movements" is emphasized by Bertalanffy. . . . These concepts represent *a complete revision of the original homeostasis principle* which emphasized exclusively the tendency toward equilibrium. [Italics partly added.]

In general, the homeostasis scheme is not applicable: a) to dynamic regulations, i.e., regulations not based upon fixed mechanisms but operating within a system functioning as a whole (e.g., regulative processes after brain lesions); b) to spontaneous activities; c) to processes whose goal is not reduction but building up of tensions; d) to processes of growth, development, creation and the like. We may also say that homeostasis is inappropriate as an explanatory principle for those human activities which are *non-utilitarian,* not serving the primary needs of self-preservation and survival and their derivatives—as is indeed the case with many "cultural" manifestations. This is so because these are based on symbolic rather than biological values (cf. Chapters 1 and 2). But even living nature is by no means merely utilitarian (Bertalanffy, 1952).

The homeostasis model is applicable in psychopathology inasmuch as non-homeostatic functions decline in mental patients. Thus Menninger, Mayman, and Pruyser (1963) were able to describe the progress of mental disease as a series of defense mechanisms, settling down at ever lower homeostatic levels until mere preservation of physiological life is left. Arieti's (1959) concept of progressive teleologic regression in schizophrenia is similar.

Differentiation

"Differentiation is transformation from a more general and homogeneous to a more special and heterogeneous condition" (Conklin, after Cowdry, 1955). "Wherever development occurs, it proceeds from a state of relative globality and lack of differentiation to a state of increasing differentiation, articulation, and hierarchic order" (Werner, 1957).

The principle of differentiation is ubiquitous in biology, e.g., the evolution and development of the nervous system, behavior, psychology and culture. Mental functions generally progress from a "syncretic" state (Werner, 1957) where percepts, motivation, feeling, imagery, symbols, concepts, etc., are an amorphous unity, toward an ever clearer distinction of these functions. In animal and a good deal of human behavior, there is a perceptual-emotive-motivational unity; perceived objects without emotional-motivational undertones are a late achievement of mature civilized man. The origins of language are obscure; but so far as we can form an idea it seems that "holophrastic" (Werner, 1957) language and thought, that is, utterances and thoughts with a broad aura of associations, preceded separation of meanings and articulate speech. Myth was the prolific chaos out of which language, magic, art, science, medicine, mores, morals, religion, etc., differentiated (Cassirer, 1953-1957).

Centralization and Related Concepts

"Organisms are not machines; but they can to a certain extent *become* machines, congeal into machines. Never completely, however; for a thoroughly mechanized organism would be incapable of reacting to the incessantly changing conditions of the outside world" (Bertalanffy, 1952). The *principle of progressive mechanization* expresses the transition from undifferentiated wholeness to higher function, made possible by specialization and "division of labor"; it also implies loss of potentialities in the components and of regulability in the whole.

Mechanization frequently leads to the establishment of "leading parts," that is, components dominating the behavior of the system. Such centers may exert "trigger causality," i.e., in contradistinction to the principle *causa aequat effectum* a small change in a leading part may, by way of *amplification mechanisms*, cause large changes in the total system. In such way, *hierarchic order* of parts or processes may be established.

In the brain as well as in mental function, centralization and hierarchic order are achieved by *stratification* (Rothacker, 1947; Lersch and Thomas, 1960; and others); that is, superimposition of higher "layers" which take the role of leading parts. Particulars and disputed points are beyond the present survey. One will agree, however, that in gross oversimplification three major layers and evolutionary steps can be distinguished. These are, in the brain, the paleencephalon in lower vertebrates; the neencephalon (cortex) progressively evolving from reptiles to mammals; and certain "highest" centers, especially the motoric speech (Broca's) region and the large association areas superimposed in man.

In some way parallel to this is the stratification observed in the mental system which can be roughly circumscribed as the domains of: a) instincts, drives, emotions, the primeval "depth personality"; b) conscious perception and voluntary action; and c) the symbolic activities characteristic of man. However, none of the available formulations, e.g., Freud's id, ego and superego and those of the German stratification theorists, is acceptable. The neurophysiological meaning of a small portion of mental processes being "conscious" is completely unknown. The Freudian unconscious or id certainly comprises only limited aspects and already pre-Freudian authors have given a much more comprehensive survey of unconscious functions (L. L. Whyte, 1960). These and related problems certainly are far from clarification.

Boundaries

Any system, as an entity which can be investigated in its own right, must have boundaries, either spatial or dynamic. Strictly speaking, spatial boundaries exist only in naive observation, and all boundaries are ultimately dynamic. One cannot exactly draw the boundaries of an atom (with valences sticking out, as it were, to attract other atoms); or of a stone (an aggregate of molecules and atoms which mostly consists of empty space, with particles at planetary-like distances); or of an organism (continually exchanging matter with its environment).

In psychology, the most important is the *ego boundary*, the distinction between self and external world, subject and objects. This distinction—with the subsidiary categories of space, time, number, causality, etc.—is slowly established in evolution and appears gradually in child development. The ego boundary differentiates from an original adualistic state (Piaget, 1959) in the human infant via phantasmic and paleologic universes (Arieti, 1964) and is probably not completely established before the *I*, *Thou* and *it* are named, i.e., before symbolic processes intervene. The latter make for the "world-openness" of man; that is, man's universe widely transcends biological bondage and even the limitations of the senses.

Symbolic activities

"Except for the immediate satisfaction of biological needs, man lives in a world not of things but of symbols" (see Chapter 1). We may also say that the various symbolic universes which distinguish human cultures from animal societies are part, and easily the most important part, of man's behavior system. It can be justly questioned whether man is a rational animal; but he certainly is a symbol-creating and symbol-dominated being throughout. For the same reason, human striving is more than self-realization; it is directed toward objective goals and realization of values (Frankl, 1959b, 1960), which means nothing else than symbolic entities which in a way may become detached from their creators (see Chapters 1 and 2). It is precisely for symbolic functions that "motives in animals will not be an adequate model for motives in man" (Allport, 1961) and that human personality is not finished at the age of three or so, as Freud's instinct theory assumed.

This is the ultimate reason why human behavior and psychology cannot be reduced to biologistic notions like restoration of homeostasis, conflict of biological drives, unsatisfactory mother-infant relationships, and the like.

10
Toward a Generalized Theoretical Model for Psychology

Necessity and Limitations of Model Conceptions

History of science shows that progress does not consist in a mere gathering of facts, but largely depends on the establishment of theoretical constructs. Idealizations never completely realized in nature, such as the conceptions of an ideal gas or an absolutely rigid body and constructs such as the structural formulae in chemistry or the planetary model of the atom, form the basis of physical theory. On the other hand, the fact that adequate model conceptions have not yet been found is the reason that many fields within the biological sciences are a mere collection of an ever-increasing amount of data, lacking exact laws and not permitting control of the phenomena in thought and in practice.

Though the necessity of theoretical models may be granted, we must be aware of their limitations also, especially as far as psychology is concerned. A first limitation can be expressed by the dictum of the scholastics: *individuum est ineffabile.* Theoretical constructs are essentially a means of establishing "laws of nature," yet all laws of nature are essentially of a statistical character. That is, they are statements about the average of a great number of events. This fact is understood in physics, where micro-events at the level of the elementary units are unpredictable in principle, whereas the seemingly deterministic laws of macrophysics result from the average behavior of a practically infinite number of elementary units. The same dictum holds true, *a fortiori*, for the higher levels of reality. We are able to state laws in the fields of biology, behavior, and sociology which are essentially laws of the average behavior of biological units considered on the cellular, organismic, and superorganismic levels.

Here, however, a peculiar situation arises. Our interest in the individual is at a minimum with physical entities, and so the statistical law gives us all the information we want. Amoebas, earthworms, and even dogs, as far as they are objects of the physiologist's research, are almost physical objects. My dog, however, and even the planaria which became familiar to me during some time of observation, are individuals. With human beings, our interest in the individual is at the maximum. It is true that we are able to establish exact laws even here for average behavior; for example, it is an empirical law that so many persons are killed per year in car accidents or are murdered. However, our interest in human beings is not satisfied by knowing just these statistical laws. We feel that another type of insight is necessary, namely, the understanding of human beings as individuals, an aspect expressed in its highest form in the work of the great artist and poet. This is the antithesis between "nomothetic" and "ideographic" attitudes, between "scientific" and "understanding" psychology (*verstehende Psychologie*); see Figure 3. Scientific psychology is

Figure 3 Diagrammatic representation of nomothetic and ideographic attitudes

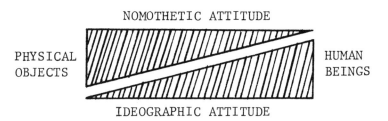

concerned with the nomothetic attitude, and it is to it that model conceptions belong.

The second limitation of model conceptions in psychology is a consequence of the fact that "inner" or "mental" experience constitutes a level of reality different from that of "outer" or "physical" experience. Our inner experience, perceptions, emotions, decisions of will cannot be reduced to action currents, hormones circulating in the blood, switching of excitations over certain synapses, and the like. The best we can hope for is to find, as far as certain aspects are concerned, a formal correspondence or isomorphy between the laws characterizing the processes in the nervous system and those found in mental phenomena. "The unity of science will not be achieved by *reducing* psychological principles to neurological ones, and neurological ones to physical ones. What we must seek is to make physical principles *congruent* with neurological ones, neurological ones with psychological ones" (Krech, 1950, p. 246).

As one bears in mind these limitations, the next step is to decide in what direction theoretical models in psychology should be sought. The situation is similar to that existing when, in the mid-1920s, an effort was made to determine, by examination of the fundamental explanatory schemes in morphogenesis, the necessary orientation in biology (Bertalanffy, 1933). Since the present author is a biologist, it appears that a demonstration of the parallelism in the modern trends of psychology and biology should be the main task of his contribution. More detailed discussions have been given elsewhere (Bertalanffy, 1948, 1952).

Actually, the number of conceptual schemes available for the interpretation of reality is rather restricted. So it is no wonder that corresponding schemes appear in different fields, such as biology and psychology, and that they often reappear within one science, "so that in many cases there is a spiral recurrence of analogous principles on more advanced levels of methodological perfection" (Brunswik, 1950, p. 75).

The main possibilities of theoretical models in psychology can be summarized in three basic alternatives which, though interconnected, can be distinguished for the purpose of analysis. These antithetic models are not necessarily mutually exclusive; rather they represent complementary and different but equally necessary approaches.

First Alternative: Molecular and Molar Models

A first alternative, the so called *molecular* vs. *molar* antithesis found in psychology may be indicated in biological terms as follows:

> For understanding the phenomena of life . . . it is not only necessary to carry on analysis as far as possible, in order to know the individual components, but it is equally necessary to know the laws of order by which parts and partial processes are integrated, and which determine just the characteristic peculiarities of life. In the discovery of these systems laws, the organismic conception sees the essential and specific object of biology. (Bertalanffy, 1952, pp. 31, 145-146)

We can either try an explanation by way of analysis into ever finer partial processes or try to establish global laws for phenomena as a whole. Functionalism emphasizes that only the latter approach leads to the essential problems and complies with the requirements of normalcy, naturalness, and "closeness to life." It appears that the way to overcome the antithesis between analysis of isolated events and global laws, between molecular and molar approaches, is to acknowledge the relative necessity of both ways:

> There is a kind of complementarity between the analytical and the system conception. We can either isolate the individual processes in the organism and define them in physico-chemical terms, whereby the whole eludes us owing to its tremendous complexity. Or we can state laws for the biological system as a whole, having to renounce, however, physico-chemical determination of the individual processes. (Bertalanffy, 1952, p. 146)

Second Alternative: Material and Formal Models

A second alternative could be termed the choice between *material* and *formal model conceptions*. We can either build hypothetical constructs in the form of assumed entities (material approach); or we can try models that are non-committal with respect to the entities concerned and only give a formalization of the laws of the phenomena under consideration (formal approach).

To the first type belong all interpretations in terms of hypothetical substance, structures, nerve connections, and the like. If such a hypothesis is correct, the entities assumed are later demonstrated in direct observation. The second way of approach is less evident though it is quite common in the evolution of science. For example, classical thermodynamics is a construct of the formal type, the notions of entropy, of the Carnot cycle, etc., being abstract and unvisualizable. Later, kinetic theory transformed thermodynamics into a theory of the material type, explaining, for example, entropy by the movement of molecules and their probable distributions. Similarly, Mendel's original system was a formal theory. It gave the laws of the distribution of hereditary characters in successive generations of hybrids, but Mendel knew nothing about chromosomes, meiosis, haploid and diploid cells, and so forth, and the material basis of heredity was discovered much later.

History of science shows that constructs of the formal type are highly useful, especially in the earlier stages of scientific development. Later, material models can be established and verified in direct observation. Adhering to material models and trying to explain all phenomena in terms of hypothetical substances or structures appeals to the human preference for what can be visualized, touched, and analyzed. It may lead, however, to the hypostatization of structures where there are none (because the order is essentially dynamic), to the adoption of one-sided elementaristic conceptions, and to the disregard or displacement into metaphysics of those problems which are not handy for material interpretation.

As far as psychology is concerned, little is known about the material counterpart of mental experience in the brain. So it may be useful, instead of elaborating hypothetical neural mechanisms, first to try a formalization of what seems to be the essential laws in this realm. This is the spirit of American functionalism whose "ignoring of the brain" has been characterized as an approach which is "less physiological and more biological" (Brunswik, 1950, p. 107).

It may be that a general theory of systems which was developed by the present author (1950b) can serve as a starting point for a "formal" approach. In fact, those very concepts which are most basic for psychological theory, such as wholeness and summativity, progressive segregation, mechanization, centralization, leading parts, finality and equifinality, anamorphosis, and so on are defined in general system theory (see Chapter 9), a framework which is ready to be filled with the contents of neurological and psychological facts.

Third Alternative:
Static and Dynamic Models

The basic characteristic encountered in biological as well as in psychological phenomena, considered from both the behavioristic and introspective standpoints, is the order and pattern of events. To explain order, there are two fundamental possibilities. The first is explanation in terms of *structural arrangements;* the second is explanation in terms of *dynamic interaction of processes.*

The static approach is, of course, represented by the classical neuron-center-localization-association theory. The pattern of the neural, and corresponding mental, processes is granted by the architecture of the nervous system. The centers represent relays or switchboards connecting incoming stimuli and excitations going out to the effectors; they are, therefore, fixed "machines" for definite functions. There is a point-to-point correspondence between, say, the excitation of elements of the retina, of the corresponding nerve cells in the visual cortex, and of elementary sensations the sum of which represents perception. Memory, association, the establishment of conditioned reflexes, etc., are based upon the building-up of nerve-paths between neurons and centers.

The criticism of classical theory as given by *Gestalt* theory need not be repeated here. It should be mentioned, however, that some important aspects of Gestalt phenomena can well be accounted for in more refined structural theories. Rashevsky indicated in 1931 a model of a thinking machine capable of gestalt discrimination, and, more recently, cybernetics has offered a new theory of neural mechanisms in general and gestalt discrimination in particular. According to Wiener (1948), *Gestalten* (that is, different perspective views of a figure recognized as the same) form a transformation group in the sense of group theory. As in ordinary television a two-dimensional plane is covered by the process of scanning, so every region in a group-space of any number of dimensions can be represented by a process of group-scanning whereby all positions in this space are traversed in a one-dimensional sequence. Such a process can serve as a method of identifying the shape of a figure independently of its size, its orientation, or other transformations, and is well adapted to mechanization. A device for group-scanning, planned as a prosthesis for the blind, was developed by McCulloch. This involved registering the *Gestalten* of printed letters by means of photoelectric cells and translating them into a series of tones of different pitch. The scheme of this array resembles the arrangement of neurons and nerve connections in the fourth layer of the visual cortex.

But there are other facts that can hardly be reconciled even with a refined machine model and that are indicative of a genuinely *dynamic order*. The first is the *principle of closure*. If incomplete figures, as for example, a circle with a little gap, are briefly flashed on a screen in a tachistoscopic experiment, movements of closure are seen; the free ends of the figure seem to join together. Or if a number of points in circular arrangement are presented with one point somewhat outside the circle, this point appears to move into the periphery in order to complete the circle according to the "law of pregnance." Phenomena of this kind, many well-known examples of which are offered in *Gestalt* theory, unmistakably show a dynamic order.

Secondly, there is a principle which is very characteristic of biological and psychological phenomena and which may be called the *principle of progressive segregation* (Bertalanffy, 1950b, 1952).

> Hierarchical order in physical systems, as, for instance, the space-lattice of a crystal, results from the union of originally separate systems of lower order, atoms in this case. In contrast, in the biological realm primary wholes segregate into subsystems. . . . Classical association psychology assumes that individual sensations, corresponding to the excitation of individual receptor elements, for example of the retina, are the primary elements, and that they are integrated into perceived shapes. However, modern research makes it probable that at first there are yet unorganized and amorphous wholes which progressively differentiate. This is shown in pathological cases. With patients recovering after cerebral injuries, it is not punctual sensations that reappear first. A point-light causes, at first, not the sensation of a luminous point, but of a vaguely circumscribed brightness; only later on, perception of shapes and finally of points is restored. Similar to embryonic development, the restoration of vision progresses from an undifferentiated to a differentiated state, and the same probably holds true for the phylogenetic evolution of perception. (1952, p. 52)

Thirdly, progressive segregation brings with it the notion of *progressive mechanization*, a principle encountered in many biological phenomena (cf. Chapter 9, p. 118).

> Primarily, organic processes are governed by the interplay within the entire system, by a dynamic order, and this is at the basis of regulability. Secondarily, progressive mechanization takes place, that is, the splitting of the originally unitary action into individual actions occurring in fixed structures. . . . The C.N.S. progresses from a less mechanized to an increasingly more mechanized state though this mechanization is never complete as shown by regulation. Phylogenetically, a progressive fixation of centers can be found in the series of vertebrates. . . . Similarly, ontogenetic investigation shows that local reflexes are not the primary element of behavior, as upheld by classical theory, but that they rather differentiate from primitive actions of the body as a whole or of larger body regions. (Bertalanffy, 1952, pp. 29, 113)

This principle may help to explain the existence of seemingly conflicting findings in experimental neurology. On the one hand, there is the vast amount of evidence upon which center and localization theory is based. The study of the responses and activities of isolated parts of the central nervous system, of the loss of functions after pathological or experimental injuries, and of localized stimulations leads to the classical picture of segmentally arranged reflex centers in the spinal cord, of reflex and automatic centers in the medulla oblongata, and of sensory, motor, and association fields in the brain. On the other hand, there is the clinical and experimental evidence for regulation, indicating the equipotentiality of the nervous system and its ability to function as a whole. Bethe's experiments (Bethe, 1931; Bethe and Fischer, 1931), for example, have shown that motoric co-ordination is re-established after amputations and thus is controlled not by pre-established central mechanisms, but rather by the entire complex of conditions present at the periphery and in the C.N.S., according to dynamic laws which have been elucidated, especially in the more recent work of von Holst (1948). Lashley's (1929) and Krech's (1950) experiments on rats as well as Goldstein's (1939) clinical observations show that localized brain injuries lead to a general deterioration of behavior and mental abilities rather than to the loss of individual functions.

Possibly the most obvious demonstration that the brain functions as a whole is "narrowness of consciousness." The fact that only one experience is in the focus of consciousness at a time seems to indicate that its physiological correlate extends over the whole "brain field." If the mosaic theory were correct, obviously any number of excitations and corresponding experiences could be co-existent.

> The theory of memory probably also must be reshaped in a similar way. Here too, the classical conception was summative and mechanistic, assuming that traits or engrams of former excitations are deposited in small groups of ganglion cells, connected by myriads of nerve connections. If, however, form perception is a system process, dynamically ordered and extended over larger brain areas, the after-effect of excitation will consist not in leaving traits in individual cells, but in an alteration of

the brainfield as a whole. Experimental and clinical facts indicate that the brain does not work as a sum of cells or sharply circumscribed centers; after brain lesions, never a single function is lost, but always others are impaired, the more the higher their demands on brain function. Thus another explanation presents itself as opposed to path theory: The process in the brain during the period of learning, when two stimuli were co-existent, represents a unitary whole; after fixation, a partial stimulus will lead to the revival of the trait as a whole, and thus to association, recognition, and conditioned reflex. (Bertalanffy, 1952, pp. 178–179; cf. also Rohracher, 1948)

Thus, the principle of progressive mechanization should provide a common denominator for the two contradicting lines of evidence and the antithetic model conceptions derived therefrom.

Mechanization is, of course, even more familiar in the behavioral and mental realms when activities that are initially plastic and under conscious control later become fixed and unconscious. This is the case in every process of learning, from the development of the child's motoric reactions to car-driving, playing the piano, and learning differential calculus. Actually, the classical explanation of learning by way of the establishment of nerve connections implies that there is at first a yet undifferentiated system where such connections can be established. Progress is possible only by mechanization; it can be achieved only by differentiation, specialization, and establishment of mechanisms that carry through the function in a fixed way and thus with minimum expense. On the other hand, this implies the fatal character of every evolution, for mechanization must be paid for by loss of versatility, and it nips other possibilities in the bud.

It is a consequence of progressive mechanization that the machine model is especially fit for the explanation of rational thinking. Reasoning according to the laws of logic and the conceptual system of mathematics is actually something like a thinking machine. We put in certain premises, the machine runs according to fixed rules, and the result drops out. Discursive thinking proceeds along a fixed path of decisions between alternatives, after the fashion of two-valued logic and the binary system applied in modern calculating machines.

This conception is much less appropriate, on the one hand, for creative thought and, on the other hand, for everyday experience and behavior, which depend on the status of the psychological system as a whole, cognition being interwoven with all sorts of co-existing perceptions and emotional and affective factors. The dependence of perception on the context of experience has been most impressively shown in the work of Ames, Cantril, and their group, and so it is no wonder that they come to "transactional" conceptions and, in their general outlook (Cantril et al., 1949), to a standpoint closely related to organismic biology. The electronic brain and the brain as a calculating machine will be able to solve problems to which the machinery was set; however, it will not be capable of autonomous re-setting, of breaking the old rules and making new ones, of inventiveness and creativeness (cf. Brunswik, 1950, pp. 134-135).

It is perhaps the profoundest objection against classical cybernetics—as it is at another level against Descartes—that "thinking" proper, and the corresponding neural mechanism, is not a primeval function, but rather a late product of evolution.

It appears, therefore, that the primary principle of neural order is to be sought in dynamics. Fixed centers, paths, and localizations are established in progressive mechanization, structural order thus gaining an ever higher significance and allowing for interpretation in terms of machine models. This is, of course, substantially the platform defended by *Gestalt* theory.

Toward an Organismic Model of Personality

Thus it appears that model conceptions in psychological theory should be (a) molar, though allowing for molecular interpretation of the individual processes; (b) formal, though allowing for future material interpretations; (c) essentially dynamic, although including structural order, established by progressive mechanization, as a derived yet most important case.

In conclusion, a tentative definition of the living organism may be mentioned: A living organism is a hierarchy of open systems maintaining itself in a steady state due to its inherent system conditions (Bertalanffy, 1952, p. 129). It appears that a corresponding definition could be applied as a general model of personality. The dynamic character of behavioral and psychological systems has already been discussed. The hierarchical organization of the processes in behavior is evident (Bertalanffy, 1952, p. 124). Hierarchical order similarly holds true in the architecture of personality. Rothacker (1947) has discussed the "stratification of personality," and it is easy to relate it to the strata of the central nervous system. Roughly speaking, three levels are superimposed. The first is the spinal cord as a reflex apparatus; the second, the paleencephalon as the organ of the depth personality with its primeval instincts, emotions, and appetites; the third, piled on top of the latter, the cortex as the organ of the day personality, the organ of consciousness.

Finally, we must acknowledge that the universe of symbols created by man's day personality distinguishes him from all other beings (Bertalanffy, 1948). It replaces the corporeal trial and error, as it is found in lower organisms, by reasoning, i.e., trial and error in conceptual symbols. Phylogenetic evolution, based upon hereditary changes, is supplanted by history, based upon the tradition of symbols. Goal-seeking behavior is a general biological characteristic; true purposiveness, however, is a privilege of man and is based upon the anticipation of the future in symbols. Instead of being a product, man becomes the creator of his environment.

On the other hand, the antagonism between the levels of personality is at the bottom of the human tragedy. If there comes a clash between the world of symbols, built up as the moral values and concepts of humanity, and biological drives, out of place in the environment of civilization, then, with respect to the individual, the situation of psychoneurosis arises. As a social factor, that universe of symbols, which is unique to man, creates the sanguinary course of history. Thus, man has to pay for his uniqueness that elevates him above other beings. Whether the levels of personality can be properly adjusted is the question upon which man's future depends.

11
Problems of Education in America

Whence Scientists?

Some years ago *Life* magazine published an editorial entitled "Can We Produce an Einstein?" observing that the achievements of modern science were made in Europe. Apparently, the earlier leaders in American science either came from or studied in Europe. De Tocqueville's famous statement was quoted that the spirit of America, though devoted to practical science, "is averse to general ideas; it does not seek theoretical discoveries." In answer to its question, *Life* referred to the fact that the United States is much more concerned with practical know-how than with theoretical know-why, as is reflected by the fact that 95 per cent of the national research budget goes for applied science, only 5 per cent for basic work. *Life*'s recommendation, therefore, was an appeal to provide larger financial resources for the latter.

Although such a recommendation is correct and intelligent, the problem is not that simple. The expectation that we shall produce Einsteins by the simple expedient of putting some additional millions or billions into basic research, bright young scientists, scientific hardware, and large research buildings will remain unfulfilled.

Since Sputnik-I was launched, there has been an enormous debate in American education, science, and research. With a gigantic increase in research and development budgets, there has still been—so far as I am able to see—no change in attitude or any reassessment of basic outlook. These past years have shown an increase rather than a decrease of the Russian lead—and by no means in the space race alone. For example, in the fall of 1960, 80 million people in the USSR were treated with oral polio vaccine, whereas in the United States it was only tested on a small scale. According to a recent book, little Ivan in the first grade is taught a vocabulary of 2,000 words, little Johnny 158 words; in the 4th grade, Ivan is prepared for literature, history, and foreign languages, but Johnny still has to babble about Mommie and Daddie.

If we reject as improbable the hypothesis that Americans are genetically stupider than other people, I believe the answer to *Life*'s question is simple: we don't want to—American universities and institutions of learning are not the place for the breeding and care of such abnormalities as outstanding scientists.

Mr. Krushchev, who was by no means a mediocre intelligence, gave this an almost classical expression. You will remember what he said after another Russian Lunik was launched: The Americans, he said, shouldn't be disappointed about Russia's conquest of space; after all, they are terribly good in inventing new tailfins.

This is precisely the point: American science excels in designing tailfins of all sorts—in diligently working out new touches, new details or convolutions of an already given body, be it the body of a car or of a theory. But it is singularly ineffective in inventing new vehicles of space or of thought.

Why is it so? The answer lies, I believe, in a degradation of the democratic dogma. It starts at the level of the elementary school when the democratic ideal of equal rights is converted into that of equal intelligence, whence the retardation of little Johnny in comparison with little Ivan. It culminates in universities and scientific production.

While there is a great hue and cry that scientists are needed and wanted, this means the need and want for trainees to work within a given framework or template of structure and organization—great hustle and bustle, lots of machines and dials to watch, aggressive publicity that so many millions are being spent and new buildings erected, headlines about what often has to be disclaimed, and so forth. It does not mean a genuine welcome for creative individuals who, by definition, are non-conformists, try something new, are sometimes awkward in public relations, less interested in quantitative expansion than in being left alone. As a matter of fact, there is a subtle borderline where achievement is penalized. While universities go desperately hunting for junior scientists and can never have enough of them, seniors are left in the lurch because, as the routine phrase goes, "Unfortunately we have no place for a scientist of your reputation, calibre, superior achievements, etc."

Leadership and Intellect

The concise expression of this is the American primadonna myth. European universities which, after all, had some six or seven centuries of experience, used to select leaders and pioneers, and it was quite common for students to come to a university not to follow a schedule for a degree, but to hear famous professors. In the United States a similar personality is apt to be labeled "primadonna"—and this is very bad indeed. This attitude misses only one detail: You just cannot hear opera without primadonnas, even if they sometimes have difficult personalities and lack of desirable togetherness; and you cannot promote science without leaders, individuals who do not fill pre-constructed moulds but make new ones.

I would like at this point to quote the following passage from the memoirs of Field Marshal Montgomery:

> My whole working creed was based on the fact that in war it is "the man" that matters. Commanders in all grades must have qualities of leadership; they must have initiative; they must have the "drive" to get things done; and they must have the character and ability which will inspire confidence in their subordinates. Above all, they must have that moral courage, that resolution, and that determination which will enable them to stand firm when the issue hangs in the balance.

Having lived through two world wars, I am not fond of brass hats and generals and I have neither the competence nor the desire to evaluate the soundness of Montgomery's doctrine from the military viewpoint. I am certain, however, that it is essentially true for science; science, to parody the Field Marshal's statement, depends on men with leadership qualities, initiative, drive, character, and moral courage. It is these qualities which the present system hampers and paralyzes.

Take, for example, the matter of scientific publications and grants for research. Roughly speaking, the principle for evaluation used to be that the scientist who established a certain reputation, by labors over many years, would be one not likely to make a fool of himself. His previous work doesn't guarantee but makes it more or less probable that his present contribution or project has merit. Our system, however, is totally different. Whether the youngest tyro or an experienced old hand is concerned, everything goes through the same big machinery, as in the stockyards of Chicago pigs of all colors and stocks are processed uniformly to make sausages.

What is the outcome of this procedure? Notwithstanding control by supposedly competent committees, our scientific and medical journals are full of superfluous, repetitive, sometimes incompetent and falsified reports. On the other hand, because of this overflood, it often takes years to have important work published—particularly if it is new and therefore causes headaches to the editorial board.

The interference with scientific productivity goes even farther. Professor H. J. Mullera has aptly defined freedom as "the condition of being able to choose and to carry out purposes.... A person is free to the extent that he has the capacity, the opportunity, and the incentive to give expression to what is in him and to develop his potentialities." I daresay this freedom is strictly limited in American science. To use Riesman's phrase, American science is "other-directed" to a hardly calculable degree—not only applied research with a prescribed practical or commercial goal, but basic science as well, controlled as it is by fashions in science and medicine, grant-giving agencies, financial considerations, and committees of all sorts, all of which often prove much stronger than the "free choice" (in Muller's terms) of the scientist.

The Group Mystique

Even more, we see a decay of academic freedom. Mind you, I am careful to keep out matters with any political implication. I do not speak of limitations of scientific communication which may have to be imposed under measures of national security and the like. But I am forced to say that not even in Hitler's Germany did I see the thought-control and censorship which appear usual at some American institutions. I had never seen before regulations such as—I quote literally—"all publications, presentations, etc., have to be cleared and approved by the research committee"—which, incidentally, was totally incompetent and only a tool of professional intrigue. I could quote examples where this regulation was made to stick and paralyzed the development of important discoveries.

The underlying philosophy of all these and many other phenomena is the mystical belief in the group, team, committee—and, I should add, exploitation of this pseudo-democratic idea for personal purposes. Of course, the group or team has an important role in science, particularly modern science with its high degree of specialization and complicated techniques. Roughly speaking, team work will be productive and indeed indispensable wherever elaboration of a given project, discovery, or idea is concerned. The group or team will never, however, replace the individual in inaugurating new developments. There is, I believe, no example in the history of science where a new breakthrough, an essentially novel discovery or theory, was the work of a group. The idea that brainstorming in a bull session will result in new revelations has no factual background.

While aware that science is but one limited sector or aspect of modern civilization, I am inclined to believe that it is what the statisticians would call a representative sample; obervation of other sectors would lead to conclusions that are different in content but consistent and parallel with those derived from science. In a certain respect, the scientist is in a favored position. Looking—as he is well entitled to do—at the cosmos as a whole and seeing *Homo sapiens* as one particular species populating a minor planet, he may well have the aloofness and mental distance to look at the human tragicomedy from a cosmic perspective.

Two Views of History

Again using the way of gross oversimplification, forced upon us by the need for brevity, there are, in principle, two well-known ways to look at the history of mankind. The one is the theory of progress, seeing in it a continuous upward movement, principally caused by an increasing control of nature. Starting with the Agricultural Revolution somewhere in the fourth millenium B.C., this movement, although interrupted by dark ages, has continually progressed in an exponential curve since the birth of modern science. We have gained nearly complete mastery of physical nature. Biological technology, such as control of disease by medicine and control of food by applied biology, is advancing. We now enter the age of psychological technology, that is, of controlling human behavior. And if we have once gained the necessary insight into the laws of society and sociological technology—that is, when education, government, and politics have become thoroughly scientific—humanity will establish the earthly paradise with, possibly the conquest of space and planetary colonization thrown into the bargain.

The other view is different. It is known as the cyclic theory of history, focusing on a sequence of entities called high cultures or civilizations. Instead of continual progress, each of them goes through a life cycle, being born at a certain time and place, growing, flowering, reaching its apex, and eventually decaying. So it was with the cultures of the past—those of Mesopotamia, Egypt, classical antiquity, China, the Aztecs, and so on. Our own period shows unmistakable symptoms of beginning or advanced decay, and if it does not die a natural death, it may even commit atomic suicide.

To the dispassionate scientist, it looks as if the cyclic model of history is the more realistic one. However, he would renounce his profession if he could not strongly protest that there has been continual and on the whole uninterrupted progress in one particular sphere of cultural activity: science and technology. But if he keeps an open mind and doesn't forget that these are but one sector of culture—that art, poetry, music, religion and even the modest aspects of customs and styles of living are just as well an expression and need of humanity as are scientific techniques, that, in our period, we have produced grand science but no grand music, sculpture, or poetry—then he cannot lightheartedly bypass the arguments of the theories of cultural cycles and decline.

This does not imply, however, cowardly acceptance of historical inevitability. History does *not* repeat itself. Patently, our civilization is different from previous ones in two all-important respects. The latter were local phenomena, whereas ours encircles the whole planet. And ours is the first consciously to realize its danger and to possess the means to control it.

And here is where education comes in. If I say "education," I mean it in the classical sense of the word as unfolding human potentialities; I do not mean it in the sense of "human engineering," handling human beings with scientific techniques for ulterior purposes. We have already alluded to the fact that, with the help of modern technique, the human animal can be engineered just as well as inanimate and subhuman nature. The technique is well known and need not be elaborated here—the control of behavior by conditioning, the use of subconscious drives and motives, of the animal instincts in the human being, and so forth. Roughly speaking, the technique was invented by Hitler for political purposes and perfected by Madison Avenue mainly for commercial ends. I believe the main objection is not in moral indignation about degradation of man as a free entity, in the unpleasant details you find in Packard's *Hidden Persuaders* and elsewhere; but rather in the fact that human and social engineering, although efficient for particular purposes of commercialism and politics and over short periods of time, is self-defeating in the long run. If you want a less highbrow formula, you may also say that you cannot fool all of the people all of the time. The persuaders work—to the profit of manufacturers of cars, refrigerators, margarine, toilet paper, etc. But, unfortunately, human nature is not completely satisfied with commercials and commodities, so people in a commercialized society are at the same time headed for mental trouble. And a society consisting merely of mechanized or engineered human beings will not survive. Even *Brave New World* needs some marionette players who themselves are above the conditioning and social engineering they impose on the others.

Individuals Reappraised

There is no miracle drug or wonder cure; this is about the most certain knowledge modern medicine and psychiatry has taught us. Nevertheless, some of the unpleasant and dangerous aspects of American life and society patently have their roots in education. It is, of course, not my intention to suggest any practical measures. But it is, I believe, within the competence of the biologist, philosopher, and social critic to recall some principles or maxims upon which practice should be based. The introduction or, rather, reintroduction of a proper scale of values is a bald necessity.

One aspect of educational philosophy which should be re-examined is the value of the human individual, together with its opposite, the theory that all individuals have equal capabilities and intelligence. This, obviously, is a parody of the American Constitution; the Founding Fathers certainly did not envisage or aspire to the manufacture of mass-men in a commercialized society. The consequence of that theory, however, is the orientation of education to fit the lowest common denominator, that is, the lowest intelligence level in the group.

Intimately connected with this is another theory fundamental in American psychology and education. It is environmentalism, the hypothesis that all individuals are born equal in their dispositions, and that only postnatal influences mould their character and mental outfit. The theory has a long history in English philosophy. It goes back to Locke's *tabula rasa* and found its classic expression by Watson, the founder of behaviorism:

> Give me a dozen healthy infants, well formed, and my own specified world to bring them up in and I'll guarantee to take anyone at random and train him to become any type of specialist I might select—doctor, lawyer, artist, merchant-chief,

and yes, even beggar and thief, regardless of his talents, penchants, tendencies, abilities, vocations, and race of his ancestors.[1]

I don't doubt that Watson is correct; only the bunch will make very poor beggars, lawyers, and doctors indeed!

In a similar vein, Skinner found "surprisingly similar performances" in organisms as diverse as the pigeon, mouse, rat, cat, and monkey. The Russians, communism notwithstanding, never fell into this trap: The pope of Russian physiology, Pavlov, reported the enormous variability of his experimental dogs in the conditioning procedure; and if genetic differences are recognized in laboratory dogs, it will be hard to deny them in humans. The existence of hereditary differences, perfectly trivial to the biologist, appears to come as a revolutionary surprise to American psychologists. Only a short while ago, I received a publication on this topic, cautiously introducing it as "a change in basic philosophy" and asking for correction should the argument be wrong.

The S-R Fallacy

A third notion requiring re-evaluation, again a biological misconception, is the so-called stimulus-response scheme. This is the idea that an organism, the human included, only responds to stimuli coming from outside and does so with maximum economy and for maintenance of its homeostatic equilibrium. In other terms, it does nothing if not stimulated or driven by maintenance needs. I have already mentioned that this theory contradicts biological fact, disregarding as it does spontaneous activities, play, and exploratory behavior. I should add that this misconceived theory has had an enormous impact on educational theory. The consequence is that child behavior essentially is conceived as "coping" with an adverse environment, and the humane educator should, therefore, make this coping as painless as possible, by reducing to a minimum any tensions and stresses imposed by scholastic requirements. Little use is made of natural curiosity and creativity, the desire for exploration, the child's inherent pleasure in activity and function. This is another background factor underlying the pitiful results of our educational system.

The S-R theory also implies a utilitarian theory of education. Of course, large parts of education are of necessity utilitarian—from the three Rs to the training of the doctor and lawyer—and they should be made even more utilitarian by eliminating, in view of the enormous extent of present knowledge, everything that is not of use for the particular purpose. But the goal of education as a whole is not utilitarian. It is not to produce mere social automata kept in adjustment and submission by conditioning; it is to produce human beings enjoying themselves in a free society. Here, again, current educational philosophy misses the point. As I have already stated, science tends to overemphasize utilitarian know-how vs. know-why; and that, in the long run, this apparently practical approach turns out to be not practical at all. This applies even more to the so-called humanities. It is precisely the definition of the cultural values of theoretical science, art, poetry, history, and so forth, that they have no short-range, utilitarian values; they are, as the Germans have it, *Selbstzweck*, goals in themselves. But for exactly this reason, they have a utilitarian value at a higher level. That poor creature called man, beset with the shortcomings of animal physiology aggravated by domestication, making his living in a continual rat race, under a thousand stresses and chased around in a complex society, transcends the image of an overburdened Pavlovian dog only by those seemingly useless, but so indispensable, realms of his more-than-animal being.

And a last precept: Smash the image of the organization man. In our discussion of science, we have already seen that the organization man also belongs to those seemingly eminently practical, but in the long run, self-frustrating ideals. I just wonder about his usefulness and desirability in industry and business, considering the higher efficiency of foreign industry—Volkswagens, Japanese transistors, Russian jets, missiles, and satellites. I am pretty sure, however, of just one point: A free society cannot be made out of yes-men. It may be eminently comfortable for business administrations when group-think, togetherness, affability, and pre-fabricated junior executives predominate, but it is equally certain that it will lead to stagnation everywhere.

On Recognizing Worth

Finally, status symbols must be replaced by human status. Democracy grants equal rights, but the pursuit of happiness implies full realization of one's *own* potentialities. Let this be recognized. Do not level down but level up. One needn't follow Napoleon's example and make generals and civil servants into fancy counts and princes, or Stalin's example and hang them with gold and silver medals like Christmas trees. But, the empty chromium symbol of the Cadillac, the snobbish admiration of European Hollywood counts can profitably be replaced by recognition of worthiness, of spiritual aristocracy, wherever they are found.

In trying to outline what the world of values is, I have discussed certain shortcomings in contemporary science. An implicit aim has been a re-evaluation of the goals of education, looking toward a needed overhauling of our value system if we are to survive in this time of trouble. Within the small compass of my presentation, I have tried to point out at least a few nooks and ribs in our educational and social structure where new timbers should replace old ones. The good old ship is still sailing even though it's overcrowded and its internal arrangement uncomfortable at times. Wondrous to see, it even grew wings and is headed for the abyss of space. But the new frontiers it is bound to reach are not outside somewhere in interplanetary space--they are so near and yet so far; they are the dimensions of the human soul.

12
On Interdisciplinary Study

Interdisciplinary Studies Are Not a New Invention

It seems to be a widespread opinion that "interdisciplinary" studies are something novel and a development of the last fifteen years or so—the prototype being some interdisciplinary project in the behavioral or social sciences with a psychologist, an economist, a psychoanalyst, etc., sitting together, supported by a large grant from the Ford Foundation. As a matter of fact, interdisciplinary work is as old as science itself and it is hardly an exaggeration to say that most of the great breakthroughs in science (and in the humanities) were "interdisciplinary," that is, transcended the borders of "disciplines" and "departments" as they existed at the time.

When Galileo founded physics, he brought together controlled observation and mathematics that were previously separated—and exactly this constituted the overthrow of Aristotelian "physics." The same happened in the humanities. When Ranke searched the archives of Vienna, Venice and Rome and founded modern historiography, he united history in the traditional sense with diplomatics, Catholic theology, study of banking operations of the papacy, and any number of other "specialties." Grimm's laws were "interdisciplinary," work transcending even the various linguistic disciplines immediately concerned; similarly in recent times, Vestris's decipherment of Linear B was possible by combination of "Greek" with odd "disciplines" such as breaking diplomatic codes. It will be hard to tell how many disciplines entered into Darwin's work as well as, in a totally different realm, into Burckhardt's *Civilization of the Renaissance in Italy.*

It would be a misunderstanding to believe that this interdisciplinary approach is a monopoly of intellectual giants, or was caused by the fact that modern specialities and departments were not yet developed; for a precisely similar process goes on up to the present day. It is just hybrid fields ranging from physical chemistry, biochemistry and biophysics to social psychology, cultural anthropology, linguistics, etc., which were the most fertile in modern science. Genetics, not long ago, was an "interdisciplinary" enterprise drawing on zoology, botany, agriculture, cytology, biochemistry, radiation physics, and a few other "specialties" and "departments." Today it is widely recognized as a separate unit, specialty and discipline. It would be trivial to belabor how many fields of the physical, biological and medical sciences are rallied into realms such as "cancer research" or "virology." At a more conceptual level, developments like information theory or cybernetics are characterized by the fact that they originate in certain fields of engineering (e.g., telecommunication, feedback systems) and widely advance into the biological and even behavioral and psychological spheres.

Whether or not fields like biophysics, genetics, general systems, comparative literature and the like are recognized as "specialties," commanding their own departments, depends to a large extent upon practical considerations and administrative reasons. Interactions between disciplines have never been so vivid and fertile as today. Warren Weaver, a co-founder of information theory and a past president of the Rockefeller Foundation and many other institutions, has admirably expressed this in a recent speech:

> About 1920 the line between chemistry and physics began to disappear. At superficial levels of application—the cookery level of chemistry and the hardware level of physics—one can still tell the two subjects apart. But fundamentally, they now have become one. Even more spectacular and surprising is the fact that biology is now in the process of becoming completely absorbed into and merged with the rest of science. The modern molecular biologist is a chemist, a physicist, a mathematician, a

sub-microscopic cytologist—in short, a *scientist*. The origin of elements, the origin of life, and the origin of species—these have now become interrelated parts of one grand problem." (Weaver, 1961)

Reasons for the Present Discontent

With all these hopeful developments going on, why is there a widespread concern about academic education and the future of science in general? The reason appears to be that, while what has been said (and infinitely more that could be said to the same effect) is commonplace among the avant garde of science, education has not kept pace with these demands, or, for that matter, with the exigencies of modern society at large.

The objections against modern "over-specialization" in science and the resulting quest for "interdisciplinary," "integrated," or "generalist" education stem from different sources and levels. It seems they can be classified under three main headings.

The first is the feeling of malaise as to the position of the specialized sciences and scholars in modern society. This is the "encapsulation" of modern man whose intellectual universe is shrinking ever more, thus leading to a meaningless or even psychopathogenic rat race. This viewpoint has been admirably discussed by Royce (1961) and therefore need not be repeated here.

A second, related viewpoint is that specialization (like patriotism, according to Nurse Cavell) is not enough. If future scientists and engineers would know a bit more about the historical bases and sociological motor forces of our civilization; if, conversely, historians would know about the biological foundations of human and all-too-human behavior, about its psychological motivation, mass psychology, the role of science in modern life and similar topics, it would go a long way toward achieving a more intelligent use of the spiritual and material resources at our disposal. It is hardly a very daring statement to say that the present world would look different if politicians—not excluding American presidents—had known a bit more about the Treaty of Versailles and even the Thirty Years

War, the geography and ethnology of Europe, the social question in the nineteenth and early twentieth centuries, foreign languages and culture, and suchlike "interdisciplinary" matters. What will democracy come to in this "Time of Troubles" if its future leaders are technicians skilled in one specialty, but otherwise remaining within the intellectual framework of teenagers with mass media, commercials, and Madison Avenue techniques providing the only source of additional information? If the universities shun this responsibility, who is going to take it? This is less of a problem under a dictatorship or in a society with a predestined ruling class (whether it is good or the contrary) but it is *the* problem in a democracy.

A third consideration is that modern specialist education apparently approaches a point of diminishing returns. The "space lag," "missile gap," and similar lags and gaps in many fields are unmistakable warning signals. If the Russians build the better missiles; if they were first to develop an oral polio vaccine; if, after many millions spent for developing fancy tranquilizers, doctors now seem inclined to prescribe the good old barbiturates; if the costly statistics upon which the cigarette–lung cancer correlation was based were recently questioned; then the American superiority is not self-evident.

One, and the customary, answer to overcome the "gaps" that have become apparent after Sputnik, is the demand for quantities: recruitment of a larger army of scientists, additional billions for research, more expensive scientific hardware, etc. In spite of the trivial fact that science demands money, this way out appears to be dubious, as may be illustrated by an (intentionally unpolitical) example. The so-called Cancer Chemotherapy Program of the United States has, in the past few years, tested thousands of chemicals, the rate being estimated at some fifty thousand per year; in 1959, thirty million dollars were spent in this program (Karnofsky, 1961). The outcome is that there are perhaps one or two dozen of chemicals which show some promise in one way or another. In a recent interview, U.S. Surgeon General Luther Terry announced "the first experience with a drug which cures cancer in man"—and this happened to be one known for years (amethopterin), effective in a very rare malignancy (choriocarcinoma). What obviously is needed are not new packs of "specialists" thoughtlessly applying a prescribed routine, but "generalists" (or whatever you like to call them) who, by introducing some new "theoretical" ideas into the field, may propose a more intelligent approach.

Notwithstanding the truism that modern research is costly, that basic research should have increased support in comparison to applied research, etc., there is also the truth that money doesn't buy everything—in particular, it doesn't buy productive and original scientific research. It seems that whether in physics, medicine, or the social sciences, the ratio between research dollars spent and results achieved tends to decrease. There is "a great plethora of $50,000 grants for $100 ideas" (Professor Gengerelli, head of the Psychology Department at UCLA). Considering such facts and the enormous amounts spent, and comparing them with scientific achievements in countries with incomparably smaller resources—Austria, the Scandinavian countries, not to forget Canada's Banting, Best and Collip—it is questionable whether the American system of graduate education is an ideal that cannot be challenged.

One must, of course, not overlook the sociological background. The European scientific community, which uncontestedly created the bases of modern science, medicine, and industry, never consisted of more than a few thousand scholars who had to go through rather stiff "initiation rituals" like the so-called habilitation for admission to academic teaching. In contrast, if an August 1961 report of the National Science Foundation stated that the number of U.S. "professional scientists and engineers" in 1960 was 1.4 million and should rise to 2.5 million in 1970, then it is obvious that the vast majority are specialized technicians (as demanded by business, armament, public health, education, social work, etc.) within a highly industrialized society. But what about the elite somewhere hidden behind these enormous numbers?

What Interdisciplinary Education Can and Cannot Do

Such are some considerations that can be brought forward in order to advocate "generalist," "interdisciplinary," or "integrated" education opening broader vistas and promises than the current "education of specialists." What can be done about it?

As a practicing scientist, I must admit that I find a good deal of muddled thinking and over-simplification in the arguments both pro and con the "broadening" of post-graduate education. This may be illustrated by a few quotations.

If Berelson (1960) poses the question whether graduate training "is to produce the men of wisdom and broad cultivation or the men of specialized skill," it doesn't make sense to me. A man with "broad cultivation" and nothing else would have made a splendid conversationalist in a society long past, but he will not find a job in industry, government or the university. A "skilled specialist" and nothing else will be quickly appointed in a variety of jobs, but it is doubtful whether he will be a productive scientist or even an inspired teacher. The world dominated by "skilled specialists" has been portrayed by social critics from Whyte's *Organization Man* to Galbraith's *Affluent Society*. Its ultimate outcome was depicted in Aldous Huxley's *Ape and Essence*, where the two Einsteins, on the leash of their subhuman masters, face each other and eventually release the bombs for mutual extermination. Developments since seem not to have disproved this vision.

Again, Berelson is asking: "Is it enough for an historian to know colonial history, or American, or modern European, or does he have to know the Middle Ages too, or the ancient world, or the Far East, or perhaps something of economics and political science?"—with the implication that this naturally leads to hopeless superficiality and "loss of depth." I can provide Berelson with a much better example. In the practice of some medical specialty, say otolaryngology or gynecology, a doctor must have training in general medicine and surgery. This presupposes more general fields such as pathology and pharmacology. Again, a fair amount of basic science is obligatory: anatomy, histology, physiology, biochemistry, which require some amount of biology and even physics as bases. Each of these "specialties" is a science of colossal extent and much harder than the specialties quoted by Berelson, because it needs not only book-learning but lots of laboratory work too.

Nevertheless, what has been outlined is not a Utopian program of "generalist" education, but the standard curriculum at every medical school. Whoever has taught in medical schools knows only too well how far the finished product remains behind the ideal. Hardly anybody will claim that the medical curriculum is ideal. Nevertheless, the pattern is standard and the question is not whether courses in a considerable number of fields lead to "loss of depth" but is along strictly pragmatic lines: How best to select data relevant for the future doctor in the various specialties; what time to assign to them; whether fields like anatomy, histology and embryology should be taught in different courses or (as is the case in some universities) as an "integrated" course; whether recent fields like biophysics and genetics should be presented in separate teaching units; and similar questions.

The Lord and every professor knows that a class of medical students is not one of mental giants, of future Pasteurs or Oslers. Nevertheless, in a field whose responsibility in taking care of the health of the nation is certainly not less than that of teaching colonial history, the system leads to passable results. And a good number (some would say too many) M.D.'s make their "specialist" contributions toward one of the many fields of medical science, as a look into the library may tell.

Again, Berelson states with some good justification that "good students are leery of taking a degree off the beaten track" because they wouldn't find "a proper berth" to lie on. However, what is "off the beaten track" only to a small degree depends on the scientific merit of the respective field, and much on administrative considerations, professors available, the market for positions, etc. Whether a Ph.D. in genetics, biophysics, history of science, comparative literature, etc., is or is not "off the beaten track" is rather a question of whether such departments and doctoral programs exist, and whether there are opportunities for people so trained.

If the above is critical of Berelson and the "pro-specialist" school of thought, I hasten to add that corresponding objections may be made against the "pro-generalist" school. For example, McGrath (1959) distinguishes "one kind of investigation," adding new facts and being "the type of research which the Germans of the nineteenth century made illustrious"; and another of "synthesis" and "conceptualization" which presumably would be promoted by "broadening education." This antithesis or dichotomy seems to be an extraordinarily strange way to characterize the achievements of Gauss, Johannes Mueller, Helmholz, Virchow, Ranke, Planck, or whatever "illustrious" name you want to insert into the equation.

Quite correctly, but somewhat trivially, McGrath asserts that the conceptualization in Einstein's relativity theory far surpasses the handiwork in conventional "factual" Ph.D. theses. Unfortunately, we find few future Einsteins in our classes, and it is dubious whether we can breed geniuses by interdisciplinary or any other sort of education. Even in the particular case of Einstein, it is highly questionable whether his achievement had anything to do with a particularly interdisciplinary education. Without looking up his biography, one is inclined to think that it hardly included non-Euclidean geometries, the philosophy of space and time, and similar interdisciplinary topics, which precisely through Einstein's work came to the fore. There is a delightful anecdote of how Einstein developed the theory of molecular movement (the classic Einstein-Smoluchowski formula) and was afterwards pleased as punch to discover that the phenomenon, under the name of Brownian movement, was known for almost a hundred years.

What such examples tend to show is that questions such as whether "radical innovations on a national scale to achieve more breadth in doctoral training are either desirable or feasible" (Berelson) or "whether graduate education ought or ought not to be limited to a special field" (McGrath) are unanswerable in any general sense. A university does not turn out "scientists" or "scholars" but doctors, teachers, physicists, biologists, linguists, agriculturists, and so on down the list. Rather, from the above considerations (which, of course, could be continued indefinitely) some observations and conclusions emerge which are pedestrian, but perhaps not useless for realistic appraisal of interdisciplinary studies and their practical implementation.

It seems that much of the futility of the argument results from the fact that the various objectives of university education are not sufficiently differentiated. The two excerpts analyzed above in some detail are rather vague about the practical goal of graduate studies, describing it as "the historian," "the psychologist," "a lifetime of independent scholarship," "training researchers or providing a valuable educational experience," breeding geniuses of Einstein's calibre, etc. Doubtless all this in some way comes under the heading of graduate education, but its practical objectives are at the same time more modest and special. A university has, first, to produce individuals who, with a sufficient capacity, fill the many needs of modern society and of which only an extremely small fraction will or even want to be "independent scholars or researchers," to use Berelson's phrase. Secondly, it should provide them with a modicum of "cultivation." Otherwise, Ph.D.'s are little more than glorified auto mechanics, commanding skill in somewhat more complex techniques and machines, but possessing, as it was sometimes formulated, only know-how, i.e., ability to apply routines, but not know-why, insight into the meaning of what they are doing. Thirdly, the university should adequately provide for the small number of gifted individuals who will eventually become original leaders in their respective fields.

For this reason, general directives whether or not graduate education should be "broadened" cannot be given. It depends on the field, the market for graduates with a diploma, the available teachers and their particular interests and abilities, the individual case of the student, and other factors.

One thing, however, is certain. Addition or "juxtaposition" of courses in different fields neither leads to "cultivation" nor to "generalist" education. The present specialization cannot be overcome by asking the science student to take some courses in the humanities and vice versa. One can hardly expect a history student to get enthused about the anatomy of the earthworm or a student of biology about the Long Parliament. To educate generalists in this way corresponds to the technique of that honorable member of the Pickwick Club who, in order to write a profound dissertation on Chinese metaphysics, looked up the headings of "China" and "metaphysics" in the *Encyclopedia Britannica*.

As we have seen, close integration of previously separate fields of science (and the same certainly applies to the humanities) is characteristic of modern developments. The student needs information about the interdisciplinary cross-relations just as about the facts, theories, techniques, etc., in his specialty. To what extent this is necessary and feasible depends upon the specialty, as well as the professional aims of the student. This is a trivial statement but perhaps not unnecessary because, in programmatic declarations, the facts are apt to be distorted. In the same way as in the previous quotations, Berelson asks: "How many fields of English literature make for 'broad' training, or is some philosophy needed too?" One need not be a specialist to answer that to the student of Elizabethan drama "philosophy" will be of small value if conceived as a curriculum from the pre-Socratics to the logical positivists. Equally certain, a deeper understanding of Shakespeare will not be attained without contemplation of the zeitgeist of the epoch of which "some philosophy" is an important aspect. I wouldn't believe that, in the concrete case, a good university teacher would be so much specialist as to forego such and similar interdisciplinary viewpoints.

The alternative "breadth or depth" is not one particular to interdisciplinary or broadened education, but equally applies to education strictly within one conventional discipline. Without being a specialist or competent in these matters, I suspect that nobody can know "all about" colonial history, psychology, or English literature (fields expressly mentioned in Berelson's study) and I am sure this isn't so in the fields I am acquainted with. It is rather an understatement that "even mature professors with a full career in their subject do not pretend to know it all" (Berelson). The present writer, for example, could easily produce many publications and testimonies which flatteringly speak of his "many different kinds" of experimental research, the "prodigious amount of material" covered in his books, or his "proficiency in many fields of science"; but any claim to be competent in more than a limited number of aspects of the fields concerned would appear to him preposterous. The only thing anybody can and must aspire to is working knowledge of a limited number of techniques, factual knowledge selected from the gargantuan cauldron of scientific knowledge, and understanding of principles. This is true irrespective of whether education and work is specialist in any field from astronomy to zoology, is interdisciplinary and comprising more than one of the conventional specialities, or is generalist with emphasis on integrating principles. Selection of what is relevant for a certain purpose, in teaching, research or practical application, is the guiding maxim (and at the same time, a difficult art) in any scientific enterprise. This—not a philatelistic collection of as many items as possible—constitutes "depth" in contrast to superficiality.

Any attempt to make an interdisciplinary or generalist program a vehicle for lowering standards should be opposed. It seems that this is the main and very justified objection against "doctor of social science" programs and the like. The present writer is not unacquainted with certain European parallels such as the Dr. Rer. Pol. (Doctor Rerum Politicarum), which is in rather low esteem. Obviously interdisciplinary programs must not be made a way towards a "quickie" doctorate which attracts precisely those students who are not able to do it the hard way in conventional study.

Finally, there is the self-evident consideration that interdisciplinary enterprises should not interfere with existing departmental structure. Desirable goals can be obtained much more securely if the existing framework is respected. Sweeping programs, which are apt to meet justified objections from the side of the specialties, are unrealistic in terms of staff and students' demands and entail financial requests that cannot be met.

Implementation of Interdisciplinary Programs

From the above considerations some realistic proposals for interdisciplinary education emerge. They keep a middle course between the conservatism of Berelson, whose attitude is essentially to maintain the status quo and who views with suspicion any broadening of training beyond the specialty, and innovations that would upset departmental and administrative organization with unproved success. These proposals are quite realistic because they are not grand plans for the future, but can be implemented any day; they do not require additional staff, facilities, outside grants, new institutes to be added to existing departments, or similar commitments.

1. *Orientation courses for students of all faculties.* The term "orientation courses" is not pretended to be the best one, but will convey the meaning. As discussed above, students of today as potential intellectual leaders of the future badly need, beside specialist training, a general "orientation" in the problems of the modern world and society. This can't be done by insertion of some science course into the humanist curriculum or vice versa. It can be done, without sacrificing "depth," by careful selection of topics.

A one-man course to this effect has been described in some detail (Bertalanffy, 1953). Seminars with a similar orientation were given under sponsorship of the Ford Foundation in southern California colleges, under titles such as "Science and Civilization," "The Nature of Institutions," "The Nature of Man," "Standards of Judgment" (see Royce, 1961). Others could be cited.

In contrast to introduction or survey courses, such courses, or seminars, should only be given by the most mature senior staff members because only they have the broad view necessary. They would be a small burden for the professors who would each have to contribute only a few lecture hours per week. Such courses should be given full credit, examinations being one paper on a topic of the course after completion.

The writer notices with satisfaction that notwithstanding the somewhat different arrangement, his own experience completely corresponds with that of the Intercollegiate Program of Graduate Studies. He stated (1953) that such courses "should be given by a practising scientist"; that it is "not a survey course in the way of introduction or general science" but "can only be taught in the senior year of college and, in more elaborate form, in post-graduate education"; and that "the program is not Utopian, and it should not be objected that it is too highbrow to be practical." Comparing this with Royce's statement (1961), complete correspondence will be found.

2. *Courses integrating several specialties.* In view of the increasing interaction between specialties and in order to min-

imize, in an already over-crowded curriculum, such duplications as often occur, integrated courses should be taken into consideration. The way in which this can be done must be obviously left to the specialties concerned. Merely as an example it may be mentioned that at some medical schools anatomy, histology, and embryology, or else biochemistry, biophysics, and physiology, are taught as one integrated course, with the effect that a much more dynamic and vivid picture of living organisms emerges. This example certainly should be taken up in other fields of study.

3. *Interdisciplinary programs leading to the M.A., M.Sc., or Ph.D. degrees.* As has been said before, the distinction between what is an interdisciplinary program or a special department is fluid. It is obvious that those fields which have not yet the status of a department should be given special attention. The main considerations would be the demand for graduates in this field, the availability of teachers, the provision of space and facilities, etc. If it makes little sense to offer programs and to make appointments when there are no takers among the students, it should, on the other hand, be clear that an authority offering an attractive program of study may well attract able students who would otherwise go elsewhere.

Some such fields coming to the mind of the present writer (a biologist) are cytochemistry and histochemistry, biophysics, bio-energetics, theoretical biology, history and philosophy of biology, general behaviorology, and general systems. Similar lists can easily be provided by representatives of other fields. None of these programs would need a large and expensive set-up; what is required is a professor, students wanting to take such programs, and a miminum of accommodation and facilities to be obtained through co-operation with existing departments.

4. *Multidisciplinary seminars with a view of educating scientific generalists.* This is the type advocated by Royce. Only two points may be re-emphasized. First, that this is a program for exceptionally gifted and enthusiastic students; second, that such seminars cannot be made to order but must grow organically from the fields concerned and common interests of the professors.

In conclusion, I wish to re-emphasize the warning that extreme care must be taken that interdepartmental Ph.D. programs should enhance the prestige of the university—as, of course, is true also of any Ph.D. thesis done in the established lines.

On the other hand, it can be stressed that if a university, and particularly a young university striving for a reputation, can offer programs otherwise not found, and if these are guided by scholars of renown, it has a rather unique opportunity that should not be missed. It is in such offerings, rather than in trying to duplicate what elsewhere is done on a larger scale, that special potentialities for a young and thriving institution exist.

Notes

Chapter 1

1. On the other hand, an uttered sound may be connected with some biological function. For example, the word for *mother* begins in nearly all languages, irrespective of their structure, with an *m*—apparently a sound connected with the smacking by the baby at the mother's breast.

Chapter 2

1. It goes without saying that the notion of progress is intimately connected with the biological notion of evolution. In geological times, the living world advanced from lower to ever higher animal and plant species. In the same manner, mankind has progressed from primitive to ever more advanced states, the peak presently attained being Western civilization of the twentieth century.

2. "The fundamental fact about American male psychology is the fear of impotence. Let's give the men, therefore, the One Big Symbol that will make them feel that they are not impotent. Let's give them great big cars, glittering all over and pointed at the ends, with 275 h.p. under the hood, so that they can feel like men!" (Hayakawa, 1957).

Chapter 5

1. Representative language (that is, possession of a "vocabulary") is a basic prerequisite for a propositional language, the combining of symbols by way of a "grammar." The latter is apparently a uniquely human feature.

2. It is not claimed that the term proposed is necessarily the best one; it is only the best the writer was able to find. We are, therefore,

inclined to retain the definition as, at the least, a working hypothesis to indicate the necessary and sufficient criteria for distinguishing between human and animal behavior.

3. Grimm's law expresses the regularities in consonant mutation; e.g., thorp-dorf; three-drei; ploug-pflug; parish-pfarre; pater-father-Vater, etc.

Chapter 6

1. [Editor's note] Von Bertalanffy's conclusions are in agreement in principle with those of Cassirer, although their starting points were different. Cassirer was a neo-Kantian, and his work was conceived as an expansion and generalization of Kant's. Von Bertalanffy began as a biologist feeling the necessity to expand the categorical framework of science, which hitherto had been unilaterally determined by physics (1933); he tried to find criteria to distinguish subhuman and human behavior (cf. Chapter 1); and was influenced by early studies on cultural relativity (1924). Von Bertalanffy developed his ideas on symbolism independently of Cassirer and only later discovered the affinity of his work with Cassirer's. It may be interesting to note that this close mental affinity, in complete independence of their respective work, also extended to quite different fields, such as the recognition of Nicholas of Cusa as a founding figure in Western philosophy, the organismic conception of life, and the relation of physics and biology.

2. Kant's Table is essentially based upon physics—more precisely, classical physics. The categories applied to biological, psychological, and sociological phenomena cannot, in Kant's manner, be disposed of by designating them as merely "regulative" whereas only the categories based on physics (mainly space, time, and causality) are truly "constitutive." As all science is a conceptual model, the models applied in the non-physical sciences have equal significance, along with the general categories of thought they imply (Bertalanffy, 1962).

3. Royce's term "sign" corresponds to our "discursive symbols," i.e., in language, mathematics, etc. He reserves the term "symbol" for what has been called here "non-discursive symbols," also including dreams, etc. These differences are of a purely semantic nature.

Chapter 7

1. The name, picture, image, part of body, and so on, is not the thing, but only a stand-in for the thing.

2. As one example from many possible ones, cf. Muller (1961): "Man in Egypt and Mesopotamia, wrote Henri Frankfort, simply did

not know an inanimate world; for him nature was not It but a living Thou. While showing some capacity for logical thought, he did not regard such thought as autonomous or necessary, and he made no distinction between subjective and objective, appearance and reality, symbol and the thing it stood for. If so, the difficulty remains that his thought was not consistently mythopoeic. All the empirical knowledge he had acquired, as in agriculture, metallurgy, and medicine, suggests that in practice he did make some distinctions. Clearly he did not depend on magic ritual alone to assure the communal welfare. His proverbs express a very practical wisdom, usually without reference to the Thou, in effect recognizing a separate human realm. His frequent anxiety intimates that whatever feeling he had of the unity of man and nature was far from a feeling of perfect community or real identity. Ritual itself implies a possible awareness that he was in some sense a stranger in the world, who unlike other animals—animals he might imitate in dances—had to take pains to keep on good terms with the Thou. In his anxiety, at any rate, he was led to reflect about the mysterious ways of his gods, seek not merely to divine but to comprehend their will. He was at least engaged in the quest of an intelligible world."

3. "In the Middle Ages both sides of human consciousness—that which was turned within and that which was turned without—lay as though dreaming or half awake beneath a common veil. The veil was woven of faith, illusion, and childish prepossession, through which the world and history were seen clad in strange hues. Man was conscious of himself only as a member of a race, people, party, family, or corporation—only through some general category. It is in Italy that this veil dissolved first; there arose an objective treatment and consideration of the State and of all things of this world, and at the same time the subjective side asserted itself with corresponding emphasis. Man became a spiritual individual, and recognized himself as such." (Burckhardt, 1960).

4. This amounts to saying that there are different "codes" in which experience is rendered. The same "given" can be represented in different languages, scripts, etc., and similarly in different *Umwelten* of animals and symbolisms of humans. There is considerable latitude in "translation" and "codes"; the limiting condition being that the code must not differ too widely from "reality" (whatever this is).

Chapter 8

1. [Editor's note] For a critique of this paper, see Lachs (1965). For a refutation of Lachs, see Bertalanffy (1966b).

2. It should be noted in passing that many so-called "private," mental data are as amenable to objective test as are physiological or physical data. This is shown by many psychological experiments. Whether an animal or human subject has the allegedly "private" experience of seeing green (or is color blind, or has another visible spectrum) can be tested by independent observers in the same way as physiological processes are tested—i.e., by observation of suitably chosen reactions to stimuli. Instruments like the polygraph permit extensive (although by no means infallible) insight into the subject's "private" mental life. Thus the commonly accepted antithesis is problematic. Experience of material things is largely "private" because it depends upon individual learning, motivation, etc.; and mental experience is largely "public" because it is verifiable by independent observers.

3. The unconscious never fit the Cartesian dualism for the excellent reason that Descartes never thought of it. Physical entities on the one hand, the conscious mind on the other—this was the Cartesian dualism; and this neat scheme was upset the moment the unconscious was discovered.

The original definition equated "mental events" with consciousness or awareness. This would make the concept of a mental unconscious self-contradictory. It is, of course, not at all difficult to say that the unconscious ultimately reduces to neurophysiological events, mnemonic traces, reverberating circuits, effects of early conditioning, coded programming, etc. But then the Cartesian problem only reappears at a deeper level. Suppose the unconscious is composed of neurophysiological memory traces; then its conversion into conscious mental processes (e.g., in the psychoanalytic interview) is just as unintelligible as is the conversion of neurophysiological events in the visual cortex into colors seen.

4. The hope of arriving at a "solution" of the mind-body problem by way of "common sense" and "analysis of ordinary language" of "the way in which we use mental and physical terms" (as was proposed by some positivists: Feigl, in Hook, 1960) is of a fantastic naïveté. We begin from the antithesis: mind-body as it has developed within Western science. However, Indogermanic languages, as well as the models we apply in scientific psychology, can express the "mental" only by physicalist similes (Bertalanffy, 1955, 1959). This is a grave handicap. Other categorizations within a different linguistic framework are quite conceivable and may permit a much more genuine and, therefore, more realistic psychology than ours. Our ordinary language and conceptual analysis would be drastically dif-

ferent if we were to start with Plato's *logistikón, thymoedés* and *epithymetikón*, with Aristotle's *animal rationalis, sensitiva* and *vegetativa*, with the *pneûma* and *psyché* of the Gnostics, with the Indian *âtman* and *karma*, or any other outlandish psychology—that is, with other conceptualizations which are not necessarily inferior to Western psychology, and may be superior.

5. One can, of course, say that a mob is a sum of individuals, that they all feel and act the same way because of exposure to the same stimuli and conditioning. However, one wonders whether this elementaristic explanation is the whole truth. Both in its sublime and bestial deeds, mass-mind seems to transcend the individuals. If, for example, the mob is excited to actions of self-sacrifice, how is even an overwhelming emotion to overcome the basic instinct of self-preservation? It is seductive to refer to terms previously introduced: In the mob, the ego barrier becomes blurred, as is also the case in pathologic states of schizophrenia or drug-induced "model psychoses." The criterion of "privacy" of mental experience would require reconsideration if emotions were indeed "infectious."

6. Even modern positivistic writing does not get away from this naive metaphysics, as is shown, e.g., by a paper by Smith (1958), approvingly quoted by Feigl (in Hook, 1960, p. 35). As in the time of Moleschott and Karl Vogt, the "living object is a swarm of particles in space"; "consciousness is a physical process or event within the living object"; and although "the exact nature of this process or event will no doubt remain obscure for a long time," "there can be little reasonable doubt of the basic fact." It is amazing how little "naturalistic" philosophers have been influenced by what has transpired in science during the past fifty years; for example, that according to quantum physics, not even so-called elementary physical events are elementary in a "swarm of particles" but show holistic characteristics of interaction; that, besides "particles," physics always contained nonmaterial entities such as energy; that Einstein's basic formula $E = mc^2$ (one very *concrete* consequence of which are atomic bombs) has eradicated the popular visual model of "particles," etc. One is hard put to understand what is meant by "consciousness being a physical process" if one does not agree with Karl Vogt that the brain secretes thought just as the kidney secretes urine.

7. What, for example, is "science"? It certainly is, so to speak, a self-propelling entity, that is, a system organized and developing according to its immanent laws. It is a "reality" in the only operational and non-metaphysical sense of the word, that is, something deeply influencing human behavior, society, life, and even survival. But it cer-

tainly is not "material," the sum of textbooks, professors, and laboratories in existence. Nor is it "mental," the aggregate of the psychologies of persons engaged in research, teaching, and administration. Also it is not a mere collective noun for certain human behavior because, as we have said, it has its own laws of systemic construction which are neither laws of physics nor of psychology. The answer, of course, is that science, like art, music, ethics, religion, and other cultural entities, is a symbolic system transcending both material things to which it may apply, and individual psychologies. It is the vice of the Cartesian dualism that it leaves no place for such entities, which are precisely those that distinguish human from animal behavior.

Chapter 11

1. Watson, 1959.

Bibliography

Akerfeldt, S. "Serological Reaction of Psychiatric Patients to N, N-dimethyl-p-phenylene-diamine." *Summaries Science Papers.* 113th Annual Meeting of the American Psychiatric Association. Chicago (1957):31–32.
Allport, G. *Pattern of Growth in Personality.* New York: Holt, Rinehart & Winston, 1961.
Arieti, S. "Schizophrenia." In S. Arieti, ed., *American Handbook of Psychiatry.* Vol. 1. New York: Basic Books, 1959.
———. "Contributions to Cognition from Psychoanalytic Theory." Address to Academy of Psychoanalysis, December 26, 1964, Montreal.
———. *The Intrapsychic Self.* New York: Basic Books, 1967.
Bavink, B. *Ergebnisse und Probleme der Naturwissenschaften.* 8th ed. Leipzig: Hirzel, 1944.
Benedict, R. *Patterns of Culture.* Boston: Houghton Mifflin Co., 1934.
Berelson, B. *Graduate Education in the United States.* New York: McGraw-Hill Book Co., 1960.
Berlyne, D. E. *Conflict, Arousal and Curiosity.* New York: McGraw-Hill Book Co., 1960.
Bertalanffy, L. von. "Einführung in Spengler's Werk." I-VI. *Literaturblatt Kölnische Zeitung,* May 1924.
———. *Kritische Theorie der Formbildung.* Berlin: Borntraeger, 1928. (English translations: *Modern Theories of Development.* Oxford: Oxford University Press, 1933; New York: Harper Torchbooks, 1962.)
———. *Modern Theories of Development.* Oxford: Oxford University Press, 1933. (Also published in New York by Harper Torchbooks in 1962.)
———. *Das Gefüge des Lebens.* Leipzig: Teubner, 1937.

_____. "Vom Sinn und der Einheit der Naturwissenschaften." *Der Student* (Vienna) 2, nos. 7-8 (1947):10-11.

_____. "Das biologische Weltbild." *III Internationale Hochschulwochen des Oesterr. College in Alpbach* (1948):251-274.

_____. "Goethe's Naturauffassung." *Atlantis* (Zurich) 8 (1949): 357-363. (English translation: "Goethe's Conception of Nature." *Main Currents in Modern Thought* 8, no. 3 [1951]:78-83.)

_____. "The Theory of Open Systems in Physics and Biology." *Science* 111 (1950a):23-29.

_____. "An Outline of General System Theory." *Br. J. Philos. Sci.* 1 (1950b):134-165.

_____. *Theoretische Biologie, II. Band, Stoffwechsel, Wachstum.* 2d ed. Bern: Francke, 1951.

_____. *Problems of Life. An Evaluation of Modern Biological Thought.* New York: John Wiley & Sons, 1952.

_____. "Philosophy of Science in Scientific Education." *Scientific Monthly* 77 (1953):233-239.

_____. "An Essay on the Relativity of Categories." *Philosophy of Science* 22 (1955):243-264. (Reprinted in: L. von Bertalanffy. *General System Theory. Foundations, Development, Applications.* New York: Braziller, 1968.)

_____. "Some Considerations on Growth in its Physical and Mental Aspects." *Merrill-Palmer Q.* 3 (1956):13-23.

_____. "The Significance of Psychotropic Drugs for a Theory of Psychosis." Mimeographed. Report to the Study Group on Ataraxics and Hallucinogenics. World Health Organization, Geneva, 1957.

_____. "Modern Concepts on Biological Adaptation." In Charles McC. Brooks and P. F. Cranefield, eds., *The Historical Development of Physiological Thought.* New York: Hafner Publishing Company, 1959.

_____. "General System Theory—A Critical Review." *General Systems Yearbook* 7 (1962):1-20.

_____. 'General System Theory and Psychiatry." In S. Arieti, ed., *American Handbook of Psychiatry.* Vol. 3, pp. 705-721. New York: Basic Books, 1966a. (Reprinted in L. von Bertalanffy. *General System Theory. Foundations, Development, Applications.* New York: Braziller, 1968.)

_____. "Mind and Body Re-examined." *Journal of Humanistic Psychology*, Vol. 6 (Fall 1966b):113-138.

_____. *General System Theory. Foundations, Development, Applications.* New York: Braziller, 1968. (London: Penguin Press, 1971. Also translated into French, German, Italian, Swedish, and Japanese.)

———. "General Systems and Psychiatry—An Overview." In W. Gray, F. Duhl, and N. Rizzo, eds., *General Systems Theory and Psychiatry*. Boston: Little, Brown, 1969.

———. "System, Symbol and the Image of Man." In I. Galdston, ed., *The Interface Between Psychiatry and Anthropology*. New York: Brunner-Mazel, 1971.

———. "Symposium on Robots, Men and Minds. Ludwig von Bertalanffy." *The Philosophy Forum* 9 (1972):301–329.

Bethe, A. "Plastizitaet und Zentrenlehre." *Handb. norm. u. pathol. Physiol.* 15, no. 2 (1931).

Bethe, A., and E. Fischer. "Anpassungsfaehigkeit (Plastizitaet) des Nervensystems." *Handb. norm. u. pathol. Physiol.* 15, no. 2 (1931).

Bindra, D., discussant. In J. H. Tanner and B. Inhelder, eds., *Discussions on Child Development*. Vol. 2. London: Tavistock, 1957.

Bleuler, E. *Dementia Praecox or the Group of Schizophrenias*. New York: International Universities Press, 1950.

Bonin, G. von. "Brain Weight and Body Weight of Mammals." *Journal General Psychology* 16 (1937):379.

Brunswik, E. "The Conceptual Framework of Psychology." *International Encyclopedia of Unified Science*, 1, no. 10. Chicago: University of Chicago Press, 1950.

Bühler, C. "Theoretical observations about life's basic tendencies." *American Journal of Psychotherapy* 13 (1959):561–581.

Bühler, K. *Die Krise der Psychologie*. Jena, Germany: Fischer, 1929.

———. *Sprachtheorie*. Jena, Germany: Fischer, 1934.

———. "Von den Sinnfunktionen der Sprachgebilde." In *Sinn and Sein*, edited by R. Wisser. Ein philosophisches Symposium. Rubingen, Germany: M. Niemeyer, 1960.

Burckhardt, J. *The Civilization of the Renaissance in Italy*. Translated by S.G.C. Middlemore. New York: Mentor Books, 1960.

Cantril, H., et al. "Psychology and Scientific Research." *Science* 110 (1949):461–464, 491–497, 517–522.

Carmichael, L., ed. *Manual of Child Psychology*. 2d ed. New York: John Wiley & Sons, 1954.

Cassirer, E. *The Individual and the Cosmos in Renaissance Philosophy*. New York: Harper Torchbooks, 1963. (Translated into German in 1927.)

———. *An Essay on Man*. New Haven, Conn.: Yale University Press, 1944.

———. *The Philosophy of Symbolic Form*. 3 vols. New Haven, Conn.: Yale University Press, 1953–1957.

Cowdry, E. *Cancer Cells*. 2d ed. Philadelphia: W. B. Saunders, 1955.

Dilthey, W. "The Dream." In H. Meyerhoff, ed., *The Philosophy of History in our Time*. Garden City, N.Y.: Doubleday Anchor Books, 1959.

Dobzhansky, T. "Human Nature as a Product of Evolution." In A. H. Maslow, ed., *New Knowledge in Human Values*. New York: Harper & Row, 1959.

Fabing, H. D. "On Going Berserk: A Neurochemical Inquiry." *Scientific Monthly* 83 (1956):232–237.

Feigl, H. "The 'Mental' and the 'Physical'." In H. Feigl, M. Scriven, and G. Maxwell, eds., *Minnesota Studies in the Philosophy of Science, Vol. 2: Concepts, Theories, and the Mind-Body Problem*. Minneapolis: University of Minnesota Press, 1958.

Frankl, V. E. *From Death-Camp to Existentialism*. Boston: Beacon Press, 1959a.

——. "Das homöostatische Prinzip und die dynamische Psychologie." *Zeitschrift fur Psychotherapie und medizinsche Psychologie* 9 (1959b): 41–47.

——. "Irrwege seelenärztlichen Denkens (Monadologismus, Potentialismus und Kaleidoskopismus)." *Der Nervenarst* 31 (1960):385–392.

——. *The Will to Meaning: Foundations and Applications of Logotherapy*. New York: World Publishing Co., 1969.

Frazer, Sir J. *The Golden Bough*. Abridged ed. New York: Macmillan, 1949.

Freud, A., and D. T. Burlingham. *Report on Hampstead Nurseries*. Cited in K. Menninger. *Love Against Hate*, New York, 1942, p. 11.

Freud, S. *The Problem of Anxiety*. New York: Psychoanalytic Quarterly Press, 1936.

——. *The Ego and the Id*. New York: British Book Center, 1952.

——. *A General Introduction to Psychoanalysis*. New York: Permabooks, 1959.

Goldstein, K. *The Organism*. New York: American Book Co., 1939.

——. *Human Nature in the Light of Psychopathology*. Cambridge, Mass.: Harvard University Press, 1940.

Gussion, P., E. D. Merle, and A. Kuna. "An Investigation of the Validity and Reliability of the Akerfeldt Test." *American Journal of Psychiatry* 115 (1958):467–468.

Hall, C.S., and G. Lindzey. *Theories of Personality*. New York: John Wiley & Sons, 1957.

Hayakawa, S. T. "Sexual Fantasy and the 1957 Car." *Etc.* 14 (1957): 163.

Hebb, D. O. *The Organization of Behavior*. New York: John Wiley & Sons, 1949.

———. "Drives and the C.N.S. (conceptual nervous system)." *Psychological Rev.* 62 (1955):243.
Henry, J. *Culture Against Man.* New York: Random House, 1963.
Herrick, C. *The Evolution of Human Nature.* New York: Harper Torchbooks, 1956.
Holst, E. von. "Vom Wesen der Ordnung im Zentralnervensystem." *Naturwissenschaften* 25 (1937):625-631, 641-647.
———. "Von der Mathematik der Nervoesen Ordnungsfunktion." *Experientia* 4 (1948):374-381.
Hook, S., ed. *Dimensions of Mind.* New York: New York University Press, 1960.
Jakobson, R. "Why 'mama' and 'papa'?" In B. Kaplan and S. Wagner, eds., *Perspectives in Psychological Theory. Essays in Honor of H. Werner.* New York: International Universities Press, 1960.
Jastrow, J. *Freud. His Dream and Sex Theories.* New York: Permabooks, 1959.
Kainz, F. *Die Sprache der Tiere.* Stuttgart, Germany: Enke, 1961.
Karnofsky, D. A. "Cancer Chemotherapeutic Agents." *C A Bull.* (American Cancer Society) 11 (1961):58-66.
Köhler, W. *The Mentality of Apes.* New York: Harcourt Brace, 1925.
Krech, D. "Dynamic Systems as Open Neurological Systems." *Psychol. Rev.* 57 (1950):345-361.
Kubie, L. S. "The Problem of Specificity in the Psychosomatic Process." In F. Deutsch, ed., *The Psychosomatic Concept in Psychoanalysis.* Monograph Series, No. 1, Boston Psychoanalytic Society. New York: International Universities Press, 1953.
Lachs, J. "Von Bertalanffy's New View." *Dialogue* 4 (1965):365-370.
Lambo, T. A. *Report to the Study Group on Ataraxics and Hallucinogenics.* Geneva: World Health Organization, 1957.
Langer, S. *Philosophy in a New Key.* Cambridge, Mass.: Harvard University Press, 1942, 1948.
Lashley, K. S. *Brain Mechanisms and Intelligence.* Chicago: University of Chicago Press, 1929.
Laszlo, E. *Introduction to Systems Philosophy.* New York: Harper Torchbooks, 1972.
Lersch, P., and H. Thomae, eds. "Persönlichkeitsforschung und Persönlichkeitstheorie," *Handbuch der Psychologie* 4. Göttingen, Germany: Hogrefe, 1960.
Levy-Bruhl, L. *Les fonctions mentales dans les sociétés inférieures.* Paris: Alcan, 1910.
Lilly, J. C. "Mental Effect of Reduction of Ordinary Levels of Physical Stimuli on Intact, Healthy Persons." *Psychiatr. Res. Rep.* 5 (1956):1-9.

Llavero, F. "Bemerkugen zu Einigen Grundfragen der Psychiatrie." *Der Nervenarzt* 28 (1957):419-420.

Lorenz, K. "Der Kumpan in der Umwelt des Vogels." *J. Ornithol.* 83 (1935):138-213. (Translated and reprinted in C. H. Schiller, ed., *Instinctive Behavior*. New York: International Universities Press, 1957.)

——. "Die angeborenen Formen moeglicher Erfahrung." *Zeitschr. Tierpsychol.* 5 (1943):235-409.

——. "Gestaltwahrnehmung als Quelle wissenschaftlicher Erkenntnis." *Z. exper. angew. Psychol.* 6 (1959):118-165.

——. "Methods of Approach to the Problems of Behavior." In *Harvey Lectures 1958-59*. New York: Academic Press, 1960: 60-103.

Luria, A. R. *The Role of Speech in the Regulation of Normal and Abnormal Behavior*. Elmsford, N.Y.: Pergamon Press, 1961.

McDonald, R. K. "Problems in Biologic Research in Schizophrenia." *Journal Chronic Diseases* 8 (1958):366-371.

McGrath, E. J. *The Graduate School and the Decline of Liberal Education*. N.Y. Bureau of Publications, Teachers College, Columbia University, 1959.

Magoun, H. *The Waking Brain*. Springfield, Ill.: Charles C. Thomas, 1958.

Maslow, A. H. "Cognition of Being in the Peak-Experiences." *J. Genetic Psychology* 94 (1959):43-66.

Menninger, K. *Love Against Hate*. New York: Harcourt Brace, 1942.

——. In *Los Angeles Times*, September 14, 1957.

Menninger, K., with H. Ellenberger, P. Pruyser, and M. Mayman. "The Unitary Concept of Mental Illness." *Bulletin Menninger Clin.* 22 (1958):4-12.

Menninger, K., with M. Mayman and P. Pruyser. *The Vital Balance*. New York: Viking Press, 1963.

Merloo, J. *The Rape of the Mind*. Cleveland, Ohio: World Publishing Co., 1956.

Minz, B. "Biochemical Aspects of Schizophrenia and Pharmacodynamic Implications," in L. Appleby, J. M. Scher, and J. Cumming, eds., *Chronic Schizophrenia*. Glencoe, Ill.: Free Press, 1960, pp. 120-148.

Muller, H. J. *Freedom in the Ancient World*. New York: Harper & Bros., 1961.

Nietzsche, F. *The Will to Power*. Translated by W. Kaufmann and R. J. Hollingdale. New York: Random House, 1967. Preface.

Opler, M. K. *Culture, Psychiatry and Human Values*. Springfield,

Ill.: Charles C. Thomas, 1956.

Oppenheimer, R. "Analogy in Science." *American Psychologist* 2 (1956):127.

Piaget, J. *The Construction of Reality in the Child.* Translated by M. Cook. New York: Basic Books, 1959.

Rapoport, A. "Utility and Application of Mathematical Models." Mimeographed. Stanford, Calif.: Center for Advanced Study in the Behavioral Sciences, 1955.

Rashevsky, N. "Learning a Property of Physical Systems. Brain Mechanisms and Their Physical Models." *J. Gen. Psychol.* 5 (1931).

Rensch, B. *Verhandlungen der deutschen zoologischen Gesellschaft* (1952):379.

———. "Trends toward progress of brains and sense organs." Cold Spring Harbor Symposium. *Quantit. Biol.* 24 (1959):291–303.

Riesman, D. *The Lonely Crowd.* New Haven, Conn.: Yale University Press, 1950.

Rosen, V. H. "Some Problems Incident to the Concept of Primary Aggression." Paper presented in Panel E, Winter Meeting of the American Psychoanalytic Association, 1956, New York City.

Rothacker, E. *Die Schicten der Personlichkeit.* 3d ed. Leipzig, Germany: Barth, 1947.

Royce, J. R. "Educating the Generalist." *Main Currents in Modern Thought* 18 (1961):99–103.

———. "Psyche, Sign, and Symbol." In J. R. Royce, ed., *Psychology and the Symbol: An Interdisciplinary Symposium.* New York: Random House, 1965.

Sandison, R. A. "The Role of Psychotropic Drugs in Group Therapy." Paper read to the Study Group on Ataractic and Hallucinogenic Drugs in Psychiatry, World Health Organization, November 4–9, 1957, Geneva.

Schachtel, E. G. *Metamorphosis.* New York: Basic Books, 1959.

Shannon, C. E., and W. Weaver. *The Mathematical Theory of Communication.* Urbana: University of Illinois Press, 1949.

Skinner, B. F. *Verbal Behavior.* New York: Appleton-Century-Crofts, 1957.

———. *Beyond Freedom and Dignity.* New York: Alfred A. Knopf, 1971.

Smith, K. "The Naturalistic Conception of Life." *American Scientist* 46, no. 4, (1958):413.

Sorokin, P. A. *Sociological Theories of Today.* New York: Harper & Row, 1966.

Spengler, O. *Der Untergang des Abendlandes.* 2 vols. 82d ed. München, Germany: C. H. Beck, 1923.
―――. *Decline of the West.* Translated by C. F. Atkinson. New York: Alfred A. Knopf, 1939.
Thurstone, E. T. *The City of Beautiful Nonsense.* New York: Dodd, Mead & Co., 1909.
Tylor, Sir. E. B. *Primitive Culture.* New York: Holt, 1874.
Uexküll, J. von. *Umwelt und Innerwelt der Tiere.* 2d ed. Berlin: Springer, 1929.
Vischer, F. T. *Auch Einer.* Stuttgart-Leipzig: Verlag Eduard Hallberger, 1879.
Waelder, R. "Critical Discussion of the Concept of an Instinct of Destruction." *Bulletin Philadelphia Assn.* 6 (1956):97–109.
Watson, J. B. *Behaviorism.* Chicago: University of Chicago Press, 1959.
Weaver, W. "Science and Complexity." *American Scientist* 36, no. 4 (1948):536–544.
Weaver, W., et al. "The Moral Un-neutrality of Science." *Science* 133 (1961):255–262.
Werner, H. *Comparative Psychology of Mental Development.* Rev. ed. New York: International Universities Press, 1957.
Werner, H., and B. Kaplan. *Symbol Formation.* New York: John Wiley & Sons, 1963.
Whorf, B. L. *Collected Papers on Metalinguistics.* Washington, D.C.: Foreign Service Institute, Department of State, 1952.
Whyte, L. L. *The Unconscious Before Freud.* New York: Basic Books, 1960.
Whyte, W. J., Jr. *Organization Man.* New York: Simon & Schuster, 1956.
Wiener, N. *Cybernetics.* New York: John Wiley & Sons, 1948.

Appendix

Evolution of the Brain

Using a crude oversimplification, which is likely to incur the well-deserved indignation of the neurologist, the neurological evolution in vertebrates can be conceived of as a superposition of three major levels of neural apparatus piled up in the course of evolution. The oldest part is the spinal cord, which is present in *Amphioxus* as a rather remote ancestral form and is essentially an apparatus of reflex response. Superimposed on the spinal cord, the brain develops in the series of vertebrates. This evolution takes place in two major steps. In accordance with the increasing development of sense organs, massive ganglia are developed at the anterior end of the spinal cord as an apparatus to react to sensory stimuli. This primeval part of the brain, known as the paleencephalon, appears almost alone in fish, but remains important in brain development even in man. Superimposed on it, the enormous neural structure of the cerebral cortex gradually develops. This so-called neencephalon is only rudimentary in fish and increases continually in the series of amphibians, reptiles, birds, mammals, and man.

In man the paleencephalon becomes the organ of primitive functions and the mediator between the cortex (the organ of consciousness) and the neurohumoral system. The cortex is the organ of the day personality, of conscious perception, feeling, and voluntary control of action and, in particular, of the symbolic activity charcteristic of man.

Now, if we survey the series of brains from lower vertebrates up to man, the characteristic is progressive cerebralization, that is, increase in the quantity and complexity of the forebrain. The increase of brain size, in particular, in the series of mammals is probably the most important example of orthogenesis—that is, evolution in a certain direction. It follows the principle of allometric growth that generally governs the increase in the size of organs with increasing body size. As an overall approximation, it appears that brain size in the series of mammals "from the mouse to the elephant" is roughly proportional to the 2/3 power of weight; that is, it follows the surface rule that is characteristic of many physiological processes. See Figure 4.

There are complications of detail not settled in an extensive and somewhat controversial literature. (For a review of recent theories on the increase of brain size, see Bertalanffy, 1951). According to a classical theory, advanced by Dubois, the increase of brain size is composed of two factors. First, so far as closely related species are concerned, brain size increases allometrically with body size. Second, if different mammalian orders are compared, there is progressive cephalization—that is, a step-wise

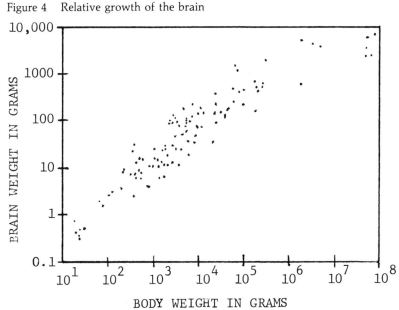

Figure 4 Relative growth of the brain

Source: G. von Bonin, "Brain Weight of Mammals," *J. Gen. Psychology* 16: 386 (1937).

doubling of brain size with the transition from a lower to a next higher group. Thus, in the series of shrew, mole, primitive and higher ungulates, anthropoids, *Pithecanthropus*, and man, brain weight (calculated for equal body weight) increases in a ratio 1:2:4:8:16:32:64. Dubois' theory is not to be regarded as proven or generally accepted, but there is little doubt that the evolution of the brain structure follows rather simple mathematical laws.

In absolute size the human brain is one of the largest, surpassed only by such giant forms as the whale and the elephant. In its relative size, or brain weight per unit of body weight, it is at the top of the series; relative brain weight is larger only in a few small mammals and birds, in which obviously brain size is correlated to the high intensity of their metabolism.

The allometric growth of the brain in the series of mammals amounts not only to a quantitative increase but, as Rensch (1952) has recently shown, further entails changes in the architecture of the brain, providing cerebral areas that can be used for higher functions. It also leads to histological changes and possibly differences in view of instinctive and learning behavior. It may be mentioned in this connection that the ventral region of the frontal lobe, where the motoric speech center is localized, is lacking even in anthropoids and appears only in man (possession of this center being connected with verbal and symbolic abilities). In contrast, the temporal lobe, where, in man, the sensory speech center is located, is rather well developed in animals. Consequently, animals like the dog, the elephant, or the horse learn to obey spoken commands and so obviously have acoustic gnosia of speech. Experiments show that this ability is lost in dogs after destruction of this region.

Index

Aggression, 23, 24, 27
  aggressive phenomena, 26–28
  "essential destructiveness"
    (Waelder) and, 24, 26, 27
  symbolic constructs and, 27–28, 30
  symbolic universe and, 25, 27
Algorithm, 4, 48
Animal symbolicum, 35, 37
Anthropomorphism, 78, 80–82
Anxiety
  consequence of symbolism, 5, 28
  existential, 10
  See also Meaninglessness; Symbolic universe, breakdown

Behavior, conditioned
  as basis of learning, 6
  breakdown of symbolic universe and, 19
  modern propaganda and, 6
  symbolism, distinguished from, 1–2, 44
Behavior, ratiomorphic (Lorenz)
  consciousness and, 70–71

Cassirer, Ernest
  concept of "I," 77
  Essay on Man, 60
  parallels Kant, 60–61

Philosophy of Symbolic Forms, 59
  symbolic forms, 59, 60–61
Categories (of experience), 97
  Kantian, 60–61, 81, 96, 160n
Centralization (of mental function), 118
Christianity
  as *ersatz* value system, 19
  Nietzsche's view of, 9, 22
Consciousness, 70–71
Crime, 12, 20
  See also Meaninglessness; Mental illness
Cybernetics
  feedback scheme and, 116
  general system theory, compared with, 116
  stimulus-response scheme and, 35

De-anthropomorphism, 80–82
Descartes, René. *See* Dualism, Cartesian
Differentiation, 117
Dualism, Cartesian, 80, 88–91, 100–102, 164
  in biology, 117
  developmental stages of, 92–94
  in introspection, 95
  in psychological theory, 87, 89, 102
  unconscious and, 162n

See also Mind-body problem
Dynamics, principles of, 127-131

Education, American, 134, 135, 136-138, 140
  interdisciplinary programs of, 155-158
  interdisciplinary study, need for, 149-155
  vs. Russian, 134
  and specialization, 147-155
Ego boundary, 38, 93
  definition, 119
  fluid, 95-96
  schizophrenia and, 38, 92
Empathy, 76, 98
Equilibrium, doctrine of (Freud), 35
Evolution, man's
  "cause" of, 71
  as creative process, 74
  symbolism and, 76

Freud, Sigmund
  equilibrium, doctrine of, 35
  man as aggressor, image of, 29
  pleasure principle, doctrine of, 15
  symbols, 58, 63, 66

General system(s) theory, 109, 111, 126
Gestalt
  in dynamic order, 131
  perception and, 94
  perception (Lorenz), 69, 70, 97
  in structural theory of psychology, 127
Grimm's Law, 53, 160n

Homeostasis, 115, 116-117

Instincts, 29
"Intrapsychic self" (Arieti), 58

Juvenile delinquency, 12, 20

Kant, Immanuel. *See* Categories, Kantian

Langer, Susanne, 43, 49
  Philosophy in a New Key, 43, 59
  vocal abilities of apes, 72, 73
Language
  as distinguishing characteristic of man, 1, 41, 71-73
  as elicitation, 44, 54
  as expression, 44, 54
  functions and types (Bühler), 45
  as representation, 44, 54, 159n, 160-161n
  as symbolic system, 1, 4, 52, 97
  transmitted by learning and tradition, 2, 45, 73
  See also Grimm's Law; Symbol; Symbolic universe; Symbolism
Lorenz, Konrad
  Gestalt perception, theory of, 69, 70, 97
  innate behavior, theories of, 23, 47
  ratiomorphic behavior, theory of, 70-71
Lysergic acid (LSD), 34, 96

Man
  as aggressor (Freud), 29
  as biological organism, 17
  evolution of, 71-72, 76
  instinctual equipment of, 29, 72
  as intrinsically active, psychophysical organism, 35, 111, 114-115
  as *Homo faber*, 42

living in symbolic universe.
  See Symbolic universe
 as rational being, 41
 social instincts of, 72
 symbolism, development of. See
  Symbolism
 See also Organism; Symbol;
  Symbolic universe;
  Symbolism
Maslow, Abraham
 cognition, theories of, 51, 52,
  99
 peak experience, concept of,
  21, 51, 99
Meaninglessness
 caused by breakdown of
  symbolic universe, 18, 20, 102
 described in modern literature,
  12
 as existential neurosis, 12, 18,
  102
 as psychopathological factor, 36
 See also Anxiety; Crime;
  Juvenile delinquency;
  Mental illness; Nihilism
Mechanization, progressive
 (mental process), 118,
 128-131
Mental illness
 breakdown of symbolic universe
  and, 20, 37, 39
 cultural framework and, 40
 meaninglessness and, 20
 scientific approaches to, 32-35
 unitary concept of, 31-33
 See also Quasi needs;
  Schizophrenia
Mind-body problem, 80, 94
 alternative conceptions of, 163n
 Erewhon and, 87
 identity, theory of, and, 90
 interaction, theory of, and, 90
 in psychiatry, 85-87
 psychophysical parallelism and,
  89

psychosomatic disorders and, 86
 See also Dualism
Model psychosis, 34, 36
Myth
 anthropomorphism of, 78
 categories of, 76, 78, 79
 as origin of symbolic system,
  76-78, 82
 as world view, 77

Neurophysiology
 psychology and, 103-104, 123
Nihilism (Nietzsche), 9, 18
 See also Meaninglessness

Open system, 112, 113
 autonomous activity of, 105,
  114-115
 creative element of, 36-37
 organism as, 36, 112, 114, 132
Organism
 as open system, 36, 112, 114,
  132

Peak experience (Maslow), 21, 51,
 99
Perspectivism, 83, 99
Physics
 culture-boundedness of, 81
 psychology and, 100-101, 104
Psychoanalysis, symbols of, 63-65
 compared with Bertalanffy's
  symbolism, 66-67
 See also Freud, symbols
Psychology
 alternative theoretical models
  of, 123-131
 dualism and, 87, 89, 102
 model conceptions, limitations
  of, 121-131
 neurophysiology and, 103-104,
  123
 physics and, 100-101, 104
 robot model and, 109, 110
 See also Dynamics, principles of

Psychology, American
  denying man's humanness, 7
  ignoring symbolism, 42–43, 58, 59

Quasi needs, 26, 28, 37

Reductionism, 15, 16, 120, 123
Reification
  as category of myth formation, 76, 78, 79
  as symptom of schizophrenia, 39, 79
Res cogitans/res extensa, 38, 89, 91, 100
  See also Dualism, Cartesian

Schizophrenia, 31, 32, 38
  See also Ego boundary; Empathy; Mental illness; Reification
Science, 16, 163n
  American, limits of, 136–137, 147, 149
"Sick society," 21
Signs, classes of, 52, 53
Stimulus-response (S-R) scheme
  education and, 141–142
  equilibrium seeking and, 35
  open system and, 106
  See also Homeostasis
Suicide, 28
Symbol(s)
  -affect interlocking, 47
  criteria of, 44–48
  definition, 2, 25, 44
  discursive, 49–52, 62
  experiential, 50
  expressive language, distinguished from, 2, 44, 45
  freely created, 44, 45, 46
  many-to-one correspondence and, 62
  nondiscursive, 49–52, 62
  one-to-many correspondence and, 62
  survival and, 46–48
  See also Freud, symbols; Language; Symbolic universe; Symbolism
Symbolic universe
  autonomous life of, 5, 25, 48
  breakdown, consequences of, 18, 19, 20, 37, 39
  as criterion of social and psychological health, 21, 40
  development of, in man, 73, 76, 77–78, 80–82
  man in, 17
  myth and, 76–78, 80–82
  See also Algorithm; Aggression; Language; Symbol; Symbolism
Symbolism
  culture-boundedness of, 80
  decisive in man's evolution, 76
  development of forebrain and, 2, 29
  origins of, 2, 63, 69, 75
  subhuman forms of behavior, distinguished from, 1, 42, 43, 44, 46
  transmitted by learning and tradition, 2, 45, 73
  See also Freud, symbols; Language; Psychology, American; Symbol; Symbolic universe
System, symbolic. *See* Symbolic universe

Umwelt, 83, 91–92
Ur-symbol (Spengler), 12

Values, human, 11, 13, 14
Verbal magic, 79
  as origin of symbolism, 2, 75